POCKET CALCULATOR SUPPLEMENT FOR CALCULUS

POCKET CALCULATOR SUPPLEMENT FOR CALCULUS

J. Barkley Rosser • Carl de Boor

University of Wisconsin–Madison

ADDISON-WESLEY PUBLISHING COMPANY

Reading, Massachusetts

Menlo Park, California • London • Amsterdam • Don Mills, Ontario • Sydney

Reproduced by Addison-Wesley from camera-ready copy prepared by the authors.

ISBN 0-201-06502-9
ABCDEFGHIJ-AL-79

PREFACE

The present text has two aims. The first aim is to use the calculator to carry out numerical experiments which illuminate points made in the calculus course. This should give the reader a better understanding of the calculus, and also a better realization of how calculus can be applied to real life problems. The second aim is to show how the techniques of calculus can be used to make the reader more efficient and more proficient in the use of the calculator, and in so doing to illustrate the wide variety of problems in which the calculator can be used to advantage.

Although the present text can be used as a supplement to the teaching of calculus with any textbook, it is particularly coordinated with "Calculus and Analytic Geometry, Fifth Edition", by George B. Thomas, Jr., and Ross L. Finney. References to the latter text are frequent. "Thomas-Finney" is condensed to "T - F", and references will appear in such forms as: "The following are Problems 1-8 at the end of Sect. 1-4 of T - F, ... ," or "T - F say ... ."

References in the present text to other places in the present text will lack any reference to a text, as: "see Sect. 2 of Chap. III."

The various chapters of this text are concerned with particular calculator topics. Any single chapter of the text is a coherent, connected whole. If all the reader desires is to learn about that topic, he or she could read the chapter straight through.

However, as indicated by the title, this text is intended to be a supplement to the calculus rather than an object of study in itself. Sometimes there is a good match between this text and the calculus text. Thus the student should read our Chap. IX, "Maxima and minima of a function of one variable," when this subject is being considered in the calculus course. And, much later, the student should read our Chap. XIII, "Maxima and minima of a function of several variables," when this subject is being considered in the calculus course.

But consider Chap. VII, "Root finding." Off and on, throughout the entire calculus course, one has to find a root from time to time. The first time will be early in the course. There the student should read some of Chap. VII, to see how the calculator can make it easier to find roots. However, some parts of the chapter have to do with root finding in situations that the student will not encounter until many months later. The student should postpone reading that part of the chapter until the time comes when those matters are being studied in the calculus course.

In order that the student can know what to read, and when, each chapter opens with a Sect. 0, "Guide for the reader." In these sections, we attempt to steer the student so that he or she will find this text a help in learning calculus, and understanding it better. And, with such coordination, the material being learned in the calculus will inevitably increase the student's skill and proficiency in using the calculator.

In the preparation of this text, we were much helped by Prof. Robert T. Moore of the University of Washington, Seattle. He read carefully and critically early drafts of many chapters, and proposed many ways to improve them. We may not have followed through on his proposals as well as we should have, but the text is much the better because of his advice, and we are glad to acknowledge our debt to him.

We are grateful for many helpful suggestions from George B. Thomas, Jr., and Ross L. Finney, the authors of the calculus text which we frequently refer to. We must say that this text lent itself well to illustration by calculator examples.

We wish to extend our thanks to Stephen H. Quigley, the Mathematics Editor for the Addison-Wesley Publishing Company, for the encouragement and cooperation he has given us.

J. Barkley Rosser

March 1979

Carl de Boor

CONTENTS

Chapter 0

A WORD ABOUT CALCULATORS

0. Guide for the reader.

This chapter is required reading. You should read it at once.

1. What kind of calculator is needed for this text?

You should have available a hand held calculator which has at least keys for the trigonometric functions (for angles in radians as well as in degrees), exponentials and logarithms, and square roots. It should also have some memory. It can be either of the reverse-polish type or of the algebraic-entry type. We shall use the abbreviations RPN and AE, respectively, for these two types. Preferably, the calculator should be programmable. We won't here explain in detail the use of any particular calculator, since we anticipate that the various readers of this text have altogether many different makes or models of calculators. We take it for granted that you will consult the manual for your own calculator to learn the various fine points peculiar to your own calculator. Presumably the instructor, or a teaching assistant, or a friend can help.

Presently there is a great number of pocket calculators on the market. In appearance, they more or less resemble each other, but they have a bewildering variety of capabilities and prices. For less than ten dollars, one can get a first class calculator that will add, subtract, multiply, and divide. Just right for a rug salesman, but totally inadequate for use with the present text.

For less than twenty-five dollars, one has a choice of several good calculators that can also handle trigonometry, logarithms and exponentials, square roots, and usually a few more things. With one of these, one could manage. Perhaps five percent of the present text might be inadequately covered, but the areas there are not crucial. However, one would find oneself spending a lot of time pressing key after key to accomplish what the programmable calculator would be doing automatically.

There are quite a number of different makes of programmable calculators that sell for a hundred dollars or less. They vary widely in quality. Some have the reputation of wearing out very fast. Others are a bit deficient in their arithmetic operations. While any of them is ostensibly suitable for use with this text, the reader might find that some would give less than complete satisfaction.

In the present text, numerical examples are worked out mainly on the Hewlett-Packard-33E, which is an RPN calculator, or the Texas Instruments Programmable 57, which is an AE calculator. We will refer to these calculators for short as the HP-33E and the TI-57, respectively. Don't expect to come up with exactly the same numerical answers on your calculator if it isn't one of these. But, you should be getting more or less the same answers, unless the example is supposed to illustrate some disastrous effect of roundoff.

Besides the HP-33E and the TI-57, which are used for the examples of the present text, programmable calculators that presently sell for less than a hundred dollars are the Sharp PC 1201, the APF Mark 90, the Casio 501P, and some others. Note that the Radio Shack EC-4000 is quite identical (except for the trade name) with the TI-57. If you can find the right discount house, you may even be able to get a Texas Instruments Programmable 58 for less than a hundred dollars.

New and improved models of calculators are coming on the market so fast that by the time you are reading this there could easily be a couple more names to add to the list above. And some of those we named may have gone out of production.

From these, one can go on up to more elaborate and expensive calculators. Their extra capabilities would rarely be of use in connection with the present text, but they would be very helpful in advanced courses in engineering and science. If you contemplate taking such courses, and can afford such a better calculator, it might be advisable to get it now and get used to it. In shopping for it, two things to ask for are indirect addressing and the capability to record a program so that you can later reestablish it in the calculator without having to key it in again, step by step.

If money is no object, you can really go overboard and get calculators with printing attachments and goodness knows what else. A printing attachment is sometimes not a bad thing to have, but it costs plenty.

2. Programming for a calculator.

To solve a problem on a calculator, one presses, or "strokes", a succession of keys on the calculator. A finite succession of key strokes is called a program.

Programs for RPN calculators differ appreciably from those for AE calculators. You should read only the explanation applicable to your make of calculator. The material devoted to these explanations will be set off by an RPN or an AE in the left margin. As an example, we now discuss the evaluation of the expression

$$\sqrt{x^2 - 1} . \tag{0.1}$$

2. Programming for a calculator

RPN

To calculate (0. 1) on an RPN calculator one would use the program:

(0. 2) x, [x²], 1, [-], [√x] .

RPN

AE

To calculate (0. 1) on an AE calculator one would use the program:

(0. 2) x, [x²], [-], 1, [=], [√x] .

AE

This sort of format for setting off the RPN and AE explanations will be followed throughout the text. Note that, since you are supposed to read only one of the two explanations above, we have used the same number, (0. 2), both for the RPN program and for the AE program.

The sequence of key strokes in Program (0. 2) is to be read from left to right. The succession of symbols has the following meaning. The first symbol is not in a box, and means that the first step is to input the value of x into the display. This is commonly done by successively pressing the keys for the digits of x in order from left to right, with the decimal point in its appropriate place. Sometimes there are special keys for constants, such as π. To input something like 3.47×10^{13}, special instructions given in the manual must be followed. Also there are assorted rules about whether one must clear the display before attempting to input a value, and other such matters. The [x²] that follows means to press the key for squaring what is in the display. This key is identified by having x^2, or something similar, engraved on it. (For some models, one must press two keys in order to square a number, such as an invert followed by a square root.) The 1 that appears means to input the digit unity into the display by pressing the unity key. We could have expressed this equivalently by writing [1] to indicate that the 1-key is to be pressed. The [-] and [=] that appear mean to press the keys so labelled. The final step, [√x] , means to press the square root key.

Consider the calculation of y from the equation

$$y = \frac{y_2(x-x_1) + y_1(x_2 - x)}{x_2 - x_1} \quad (0.3)$$

Even fairly inexpensive nonprogrammable calculators usually have the capability to store a number, which can subsequently be called into the display as often as required. Let us assume at least this memory capacity. If your calculator does not have at least this, it is probably not really suitable for use with this text.

RPN

To calculate y by (0. 3) on an RPN calculator, one could use the program:

RPN

RPN

(0. 4) x, [STO] x, x_1, [−], y_2, [×],
x_2, [RCL] x, [−], y_1, [×], [+],
x_2, [↑], x_1, [−], [÷].

RPN

AE

To calculate y by (0. 3) on an AE calculator, one could use the program:

(0. 4) x, [STO] x, [−], x_1, [=], [×], y_2,
[+], [(], y_1, [×], [(], x_2, [−], [RCL] x, [=],
[÷], [(], x_2, [−], x_1, [=].

We are here assuming a parenthesis capability for the calculator. If your calculator does not have this capability, you will perhaps have to resort to calculating $y_2(x - x_1)$, storing it, then calculating $y_1(x_2 - x)$, then adding this to the stored term, and finally calculating $x_2 - x_1$ and dividing it into the stored sum (or else multiplying by $(x_2 - x_1)^{-1}$). This would be pretty clumsy, and probably would involve inputting x twice.

Some AE calculators, for example the TI-57 and other Texas Instruments calculators, are hierarchical in their arithmetic. This is part of what they call their AOS system. This means that they will carry out Multiplication and Division before Addition and Subtraction (some people remember this by thinking of "My Dear Aunt Sally"). For example, the program

1, [+], 2, [×], 3, [=]

will give the result $7 = 1 + (2 \times 3)$ on a hierarchical calculator, while a calculator that isn't hierarchical would give $9 = (1 + 2) \times 3$ for the answer. On a hierarchical calculator, there is no need for the first [(] in (0. 4) before the y_1, since y_1 is first multiplied by the parenthetical expression $(x_2 - x)$ following it before the addition to the expression $(x - x_1)y_2$ preceding it is carried out.

Finally, we have assumed that the = key finishes off all pending parenthetical expressions. For this reason, there are no closing parentheses in Program (0. 4). In the unlikely case that your AE calculator does not have this feature, or in case you want to find out the value of these parenthetical expressions, you would have to insert two [)]'s before the second [=] and one [)] before the last [=].

AE

In Program (0. 4), we have used [STO] x and [RCL] x to indicate respectively storing x somewhere and then recalling it at the appropriate time.

We shall see in Sect. 3 of Chap. VI that there may be cases when, for a fixed x_1, y_1, x_2, y_2, one will wish to calculate y by (0. 3) for several different x's. With Program (0. 4), this will involve a lot of inputting of x_i's and y_i's. This involves danger of error, as we shall explain shortly. As they are the same x_i's and y_i's each time, there must be ways to circumvent some of this by storing some of these numbers and some intermediate results in memory. Of course, this requires a calculator with more than one memory location. For such a calculator, it is customary to refer to the various memory locations as "registers." We will use "register n" for the location that one gets into or out of by pressing the digit n on the calculator.

Let us make some step by step improvements in Program (0. 4).

RPN

Actually, if we have an HP-33E in hand, we would not use Program (0. 4), but possibly something like:

(0. 5) x, [STO 0], x_1, [-], [STO 5], y_2, [×], x_2, [RCL 0], [-], [STO + 5], y_1, [×], [+], [RCL 5], [÷].

Here we use [STO 0] as an abbreviation for the two-stroke sequence [STO], [0]; it means to press the STO key followed by the 0 key. This has the effect of storing whatever is in the display into register zero. Similarly, [RCL 0] stands for [RCL], [0] and means to press the RCL key followed by the 0 key. This has the effect of reading out whatever is in register zero into the display. [STO + 5] stands for the three-stroke sequence [STO], [+], [5]. Its execution has the effect of adding whatever is in the display to whatever is in the register five and leaving the sum in that register.

RPN

AE

Actually, if we have a TI-57 in hand, we would not use Program (0. 4), but possibly something like:

(0. 5) x, [STO 0], [-], x_1, [=], [STO 5], [×], y_2, [+], y_1, [×], [(], x_2, [-], [RCL 0], [)], [SUM 5], [=], [÷], [RCL 5], [=].

Here we use [STO 0] as an abbreviation for the two stroke sequence [STO], [0]; it means to press the STO key followed by the 0 key. This has the effect of storing whatever is in the display into register zero. Similarly, [RCL 0] stands for [RCL], [0] and means to press the RCL key followed by the 0 key. This has the effect of reading out whatever is in register zero into the display. [SUM 5] stands for [SUM], [5]. Its execution has the effect of adding whatever is in the display to whatever is in register five and leaving the sum in that register.

AE

Though better is to come, we have already cut down the inputs as compared to Program (0. 4). In the process, we have made use of <u>register arithmetic</u>. By this we mean the capability of adding (or subtracting) the number in the display directly to (or from) the number in some register, leaving the result in the same register. In subsequent programs, we will in the same way multiply a number in some register n by what's in the display or divide what's in the display into the number in some register n, leaving the result again in that same register n.

The actual key sequence to accomplish such register arithmetic changes from calculator to calculator. For example, in order to divide into register 3, the HP-33E requires the key sequence

[STO], [÷], [3],

while the TI-57 requires the sequence

[INV], [2nd], [SUM], [3].

For simplicity and uniformity, we use in this text instead the symbolic keys

[STO + n], [STO -n], [STO × n], [STO ÷ n]

to indicate that the number N in register n is to be replaced there by the number $N+x$, $N-x$, $N \times x$, and N/x, respectively, with x the number currently in the display.

You should make sure you understand how the programs given above work.

If you wish to perform a calculation that requires more than a few strokes, you definitely should write out the program beforehand. This is particularly urgent if the program is to be used more than once. But it is a good habit to get into even for programs which are intended to be used only once. Otherwise it is very easy to lose your place in a calculation.

If one frequently has to use a program more than once, it is advisable to acquire a programmable calculator. A programmable calculator allows you to store an entire program as a package. After you press the proper program key (or keys), the calculator will then repeat the entire program, step by step. Thus, in Program (0. 2) you could store all the instructions after the first, in what is called "the program or "the program memory". If at some time you wish to calculate (0. 1) for some x, you would input the x, and then press the proper program key (or keys). The calculator would do all the rest automatically, and out would come a value for

$$\sqrt{x^2 - 1}\ .$$

If you wish to do this for a second x, simply input the new x, press the program key (or keys) again, and out comes the new answer. Quite a saving of time and labor, not to mention that you do not risk making a mistake while keying through the steps of the calculation.

If you do not own a programmable calculator, you can still carry out most of the calculations for the present text. You will have to run through the succession of key strokes by hand every time a program appears instead of being able to call forth the entire program by pressing one or two program keys. Use of a programmable calculator with the present text will considerably curtail the labor of calculating. However, to reiterate, you can mostly manage with a calculator that is not programmable, but you will have to take more time for the calculations.

While use of a programmable calculator will save time with Program (0. 2), it will save much more time with Program (0. 4). One of the inconveniences of Program (0. 4) is the six inputs of x_i's and y_i's. This is particularly annoying if you have to run Program (0. 4) several times for different x's but with the same x_1, y_1, x_2 and y_2. If the x_i's and y_i's are 10-digit numbers, as they could well be, all this inputting could be quite a chore, not to mention a considerable risk of pressing a wrong digit occasionally.

Let us now see what happens with a programmable calculator. We will store all of Program (0. 4) except inputting x at the beginning. This leaves us the option of running Program (0. 4) several times with different x's by just inputting a new x each time. One puts the calculator into store program mode, and then just keys through Program (0. 4) (omitting the first step) exactly as if one were performing the calculation. No calculating is done, of course, because the calculator is busy storing the program. When one comes to input x_1, one will successively press the keys for the digits of x_1, from left to right. All this will be recorded, so that when the calculator repeats the program, x_1 will automatically be input by the calculator. The same for y_1, x_2, and y_2, in their turn. So there is no more bother about having to input the x_i's and y_i's if one runs the program for several different x's. The calculator does it all for you.

However, there is still a difficulty. When one is inputting an x_i or a y_i, each digit uses up a register in the program memory. If an x_i or y_i has 10 digits, it will use up 10 program registers. There are six such inputs in Program (0. 4). If each of the x_i's and y_i's has 10 digits, we would use up 60 program registers just to store the digits. This is more program registers than are available on either of the suggested calculators.

We could go to Program (0. 5), where we have reduced the number of inputs to four. However, we could still be in trouble. Suppose x_1 is a 10-digit negative number in scientific notation with a negative coefficient. What with recording two pressings of the change sign key, recording the exponent, etc., x_1 could use up 15 program registers. Four inputs could run to 60 program registers, which are more than are available on either of the suggested calculators.

We can still manage. Store x_1, y_1, x_2, and y_2 respectively in memory registers one, two, three, and four. Then rewrite Program (0. 5) by replacing the inputs of x_1, y_1, x_2, and y_2 respectively by [RCL 1], [RCL 2], [RCL 3], and [RCL 4]. One could similarly rewrite Program (0. 4) if one chooses to store x in register zero.

Now Programs (0. 4) and (0. 5) are short enough to fit very comfortably in either of the suggested calculators. In fact, for both calculators, though [RCL 1] means to press RCL followed by 1, these two key strokes are "merged" in recording the program, so that [RCL 1] takes up only a single program register. Indeed, some three-stroke sequences, such as [STO + 5] or [INV SUM 5], are "merged" into a single program register.

The steps of a program are stored in numbered program registers. On the two suggested calculators, these program registers are identified by two digits. (In particular, register number one is called 01.) When you get ready to run a program, you may not be at the beginning of the program. You can press a combination involving GTO to get to the beginning (see your manual for details; other ways to get to the beginning are often available). Then you press a key to make the program run.

So, if you have the x_i's and y_i's stored in memory registers, and the modified Program (0. 4) or (0. 5) stored in the program registers, and wish to calculate y for a given x from (0. 3), you proceed as follows. Input x, press a combination to get to the beginning of the program, press the key to run the program, sit back and relax, and quickly y will appear.

You can add a suitable instruction to the end of Program (0. 4) or (0. 5) to take care of getting you back to the beginning, so that if you then wish to calculate y for a different x all you have to do is to input the new x and press the key to run the program.

Now that we have finished explaining how to program the calculation indicated in (0. 3), let us point out that this was altogether the wrong way to do it. We went through all this because it seemed particularly suitable for elucidating some of the high points about writing and storing programs. However, it would have been much more efficient just to point out that the equation

$$y = y_1 + \frac{y_2 - y_1}{x_2 - x_1}(x - x_1) \tag{0. 6}$$

gives exactly the same value of y as does (0. 3). But the right side of (0. 6) is far easier to program than the right side of (0. 3). The whole matter is discussed in detail in Sect. 3 of Chap. VI, where (0. 6) appears as equation (3. 3). There everything proceeds much more easily than our treatment of (0. 3).

Moral. A calculator cannot take the place of mathematical skill; not even a programmable calculator. That is why you signed up for this calculus course, in order to acquire additional mathematical skills. What has calculus got to do with replacing (0. 3) with (0. 6)? As will be pointed out in Sect. 3 of Chap. VI, (0. 6) involves the calculus notion of slope of a line. This concept is central in doing the programming for the right side of (0. 6).

Incidentally, (0. 3) and the equivalent formula (0. 6) are not something that we just made up. One or the other of them was probably taught to you as the formula for interpolating in a table. They are much used in mathematics, and are widely used in scientific calculations.

3. Errors.

A very common error in using a calculator occurs while inputting a number. It is very easy to get a digit wrong, especially if it is a 10 digit number. It is not unusual, when inputting a 10 digit number, to interchange two digits, or to leave one out, or to press one twice. One should develop the habit of checking the number in the display after inputting and before proceeding further. One should also try to write programs that call for inputting numbers as seldom as possible. If your calculator has an adequate supply of memory registers, it is good practice to store some intermediate results in them for later recall. Of course, this requires you to write out the program beforehand in order to keep a record of what is where.

One can also make errors in copying out numbers from the calculator display. The same remarks apply to this as to inputting numbers. Of course, one has to write the final answer, but one should avoid intermediate write outs as far as possible. One should develop the habit of checking the number after copying it from the display and before proceeding further.

Start developing these habits right away and persist until they are thoroughly ingrained.

Another source of error is to get the program wrong. As we said before, it is very advisable to write out each program before using it. And you should develop the habit of checking a program after it is written and before using it. Often, you know the answer to your problem for certain special choices of the input parameters. In that case, try the program you wrote for the problem on this special input to see if it gives the known answer.

In writing numbers in the present text, we choose to group five digits together and then leave a space before the next set of five (or fewer) digits. Sometimes, we will use the decimal point as a divider between groups of digits. Thus we will write an approximation for $\sqrt{2}$ as 1.4142 13562. Some people would write this as 1.414,213,562. And indeed some of the Hewlett-Packard calculators would show the number this way in the display.

It is customary to make a distinction between precision and accuracy. <u>Precision</u> has to do with the number of significant digits given. The more significant digits that are shown, the greater the precision. There has to be an understanding about which digits are significant. Thus, 1.4142 obviously has five significant digits. It is just as precise as 1.4142×10^{18}, which also has five significant digits. It is more precise than 1.41×10^{-18}, which has only three significant digits. All three of the above are less precise than 1.4142 13562, which has ten significant digits. However, if we write a number as

$$1.4142\ \ 00000 \times 10^{18}$$

or

$$1.4142\ \ 00000\ ,$$

the fact that we wrote the extra zeros (which otherwise would be superfluous) signifies that they are to be considered as significant. Both numbers have ten digit precision, in spite of the unusual fact that the last five digits happen to come out to be zero.

Calculators are erratic in their treatment of significant final zeros. Sometimes, significant final zeros will not be shown in the display. In some cases there will be blanks where there should be zeros, and in other cases the number is moved to the right, leaving blanks on the left. Alternatively, a string of zeros may be shown at the end in the display even if they are not significant.

For the numbers shown in this text, final zeros will be written only if they are significant, except when we are citing a display on a calculator. If the calculator shows non-significant zeros at the end, we may just copy what shows in the display (and try to make it clear to the reader that we are just copying).

Accuracy has to do with how far off a cited number is from what would have been the result in an ideally perfect calculation; the latter is to be considered the "correct" value. If N is a correct value and n is the number that is obtained, then $N - n$ is called the absolute error and

$$\frac{N - n}{N}$$

is called the relative error (unless $N = 0$, in which case the formula has no meaning). Suppose someone gives the estimate

$$2.7182\ 81828\ 45904\ 5$$

for π. This estimate is very precise, indeed precise to 16 significant digits. However, it is not nearly that accurate as an estimate. The absolute error is about 0.42 and the relative error is about 0.13, or 13%. The much less precise estimate, 3, would be more accurate, with an absolute error of about 0.14 and a relative error of about 0.045, or 4.5%.

In giving a numerical value for something, we have tried to be careful to use $=$ only when we have been able to get the value exactly. This includes the situation in which we only write down the first few significant digits of a number and indicate the omission of the subsequent digits by the elision sign "...". Otherwise, we use $\cong$. Thus, we will write

$$\sqrt{0.04} = 0.2$$

and

$$\sqrt{2} = 1.4142\ 13562\ 37309\ 5048\ldots,$$

but

$$\sqrt{2} \cong 1.4142\ 13562\ 37310\ .$$

Chapter I

FUNCTIONS

0. Guide for the reader.

Functions are very important in calculus. Usually the subject will be brought up in the first week or two of the course. At that point, the reader should read this entire Chap. I, which is mostly descriptive and has to do with means of calculating functions. After some discussion of functions in general, the calculus turns to a study of what can be done with, and to, functions. A second reading of Chap. I is probably advisable then. Some points should come through more clearly at that time.

From then on, through most of the calculus course, attention is confined almost exclusively to functions which can be represented by formulas. Toward the end, your calculus course may touch on the topic of differential equations. In engineering, and the experimental sciences, most differential equations have for their solution a function which cannot be represented by a formula.

If the calculus course does touch on differential equations, the reader should then broaden his understanding of functions by reading the last two sections of this Chap. I yet a third time.

1. Classical definition.

Calculus is mainly a study of functions. The notion of function is explained very early in the calculus text (see, for example, Sect. 1-6 of T-F). You should be sure you really understand it. Indeed it would not hurt if we reiterate it herewith. For f to be a function, f must be a rule, or procedure, or mechanism which for a given x (in the domain of f) prescribes a unique y, called $f(x)$.

Throughout much of calculus, and almost universally in the engineering and scientific uses of calculus, the main question about some function f is: "Given a numerical value of x, what is the unique corresponding numerical value $y = f(x)$?"

Why should this be a problem? For f to be a function, it must prescribe y, and uniquely so. But there are ways of prescribing y that do not give much of a clue as to how to calculate a numerical approximation for its value. Thus y may be prescribed as the unique root of some equation. (See Chap. VII.)

For an extreme case, let f(t) be the temperature at time t at the Greenwich Observatory. Certainly, if a person is standing at the Greenwich Observatory with a thermometer, it will register one and only one temperature at any time t. So, f is unquestionably a function. But who would presume to try to calculate a numerical approximation for what the temperature was at the time when John Hancock was signing his name to the Declaration of Independence?

However, for most functions f that the reader will encounter in calculus, there are means to calculate a numerical approximation for $y = f(x)$. Indeed, a fair amount of the calculus course is concerned with developing various such methods. The advent of the calculator is a great help with such calculations. For the rest of the chapter, we summarize the main ways to calculate an approximate numerical value for $y = f(x)$.

2. As approximated by calculator keys.

On any given pocket calculator, certain functions can be approximately evaluated by pressing a single key (sometimes two keys). You should study the manual for your calculator to learn what functions are thus directly available.

Typically, one might wish to evaluate $\sqrt{x}$, for some numerical value of x. Recall that x must be in the domain of $\sqrt{\ }$; that is, the value of x cannot be negative. First one inputs x into the display. Then, with x in the display, one presses the key labeled $\sqrt{x}$ and an approximation for $\sqrt{x}$ will appear in the display.

This number in the display, representing $\sqrt{x}$, is unique; it will come out the same as long as x is the same (unless something has gone wrong with the calculator). So, in accordance with the classical definition of function, the $\sqrt{x}$ key is a function. In accordance with the classical definition, it will work only if the independent variable, x, is in the domain of the $\sqrt{x}$ key. If x is taken negative in the display, then pressing the $\sqrt{x}$ key will cause the calculator to indicate an error, and to cease operating properly. The manual tells how to get it operating properly again.

Although the $\sqrt{x}$ key is unquestionably a function, it is not exactly the calculus function "square root of", though the calculator manufacturers try to make it very close. A difficulty is that the display can show only a certain number of digits; often 10 digits. With only 10 digits, one cannot display the exact value of $\sqrt{2}$. The closest one can come is 1. 4142 13562. The good calculators will give this number (or perhaps eight digits of it) if $\sqrt{2}$ is called for. See Chap. IV for a discussion of this source (and other sources) of inaccuracy in the use of a calculator.

Problem 2. 1. See what your calculator gives for $\sqrt{2}$, $\sqrt{20}$, and $\sqrt{200}$.

Problem 2. 2. See what your calculator gives for $\sin(1/2)$, $\sin 2$, $\cos(\pi/4)$, and $\cos \pi$.

NOTE ESPECIALLY. For use with calculus, angles MUST be taken in RADIANS. There is a special setting on the calculator for this. Do NOT forget to set this to radians

2. As approximated by calculator keys

whenever dealing with trigonometric functions.

Problem 2. 3. Calculate $10^{20} \times 10^{20}$, $10^{20} \times 10^{20} \times 10^{20}$, $10^{20} \times 10^{20} \times 10^{20} \times 10^{20}$, etc., until the calculator refuses to give an answer and shows an overflow condition. Learn about this from the manual.

The function represented by the $\sqrt{x}$ key differs from the calculus function "square root of" in yet another way because it has a different domain. For the calculus function, the domain consists of all nonnegative real numbers. But suppose that on a calculator one wishes to take the square root of

$$0.6931\ 47180\ 55994 .$$

First one has to input this into the display. But it has too many digits. If one has a 10 digit display, the closest one can come to inputting the number above is to input

$$6.9314\ 71806 \times 10^{-1} .$$

In other words, the number given earlier is not in the domain of the function represented by the $\sqrt{x}$ key, even though it is a positive real number. There are things that a calculator cannot do, and the reader must learn not to expect too much of it.

Besides the functions that are given on the reader's calculator by a single key stroke, many others can be made by combining several key strokes into a program. For example, suppose f is defined so that

$$f(x) = \sqrt{x^2 - 1} \quad \text{for} \quad |x| \geq 1 . \tag{2.1}$$

(For a definition of $|x|$, the absolute value of x, see Example 14 in Sect. 1-6 of T-F.) One would carry out the evaluation of f by using the following program:

RPN
|
RPN

x, [x^2], 1, [−], [$\sqrt{x}$] .

AE
|
AE

x, [x^2], [−], 1, [=], [$\sqrt{x}$] .

Manufacturers of calculators have exercised considerable ingenuity in arranging for the user to be able to put together simple functions to get a more complicated one. The manuals discuss many examples of such more complicated functions that can be built up out of a succession of individual key strokes; that is, functions that can be evaluated by a program. And, of course, the programmable calculators let one carry out the entire program by pressing the appropriate program key (or keys).

Problem 2. 4. Give a program for evaluating f if $f(x)$ is defined by:

(a) $f(x) = \dfrac{\sin (x^2)}{\cos^2 x + 1}$

(b) $f(x) = \sqrt{(x-1)/(x+1)}$

(c) $f(x) = \begin{cases} x & \text{when } x \leq 1, \\ 2-x & \text{when } 1 < x. \end{cases}$

What is not mentioned in the manuals is the fact that many functions cannot be given by any program whatsoever. Finding roots of an equation often falls into this category. (See Chap. VII.) However, by using the ideas of calculus, one can usually find a program whose execution will bring one very close. As a matter of fact, the various functions represented by keys on a calculator are nothing more than built-in programs designed to produce reasonably accurate approximations for the function values. So, with calculus to tell us which keys to press, and the calculator to perform the calculations quickly, we can now handle functions that used to be almost inaccessible.

3. As approximated by a table.

It has been traditional to give representations of functions by means of tables. You may recall the tables of trigonometric functions and of logarithms from your trigonometry text. Three small tables are given at the very end of T- F. In Table 3. 1, we reproduce their Table 2 from p. A-21.

In the first place, the domain of x is very limited, consisting only of the values in the first and fourth columns. However, for each x in the first column, a unique value is given for e^x in the second column and another unique value is given for e^{-x} in the third column. Similarly for the next three columns. So we certainly have defined two functions, which are quite similar to the calculus functions "exponential of" and "reciprocal exponential of". Although the table does not pretend to give more than approximate values of e^x and e^{-x}, it and similar tables have been accepted for hundreds of years as the main way to present functions. Starting somewhat later, the slide rule was also widely used to present certain functions with very limited accuracy.

One might think that the advent of the small electronic calculator would put an end to the dissemination of tables. Certainly, for the reader with a calculator in hand, the three tables of T- F are quite superfluous. However, we refer the reader to Chap. XV . There we discuss the solution of differential equations. There are cases in which the desired solution is a function that cannot be given by any program. Nevertheless, we shall learn in Chap. XV how to make a table of approximate values of the function which is the solution. So then the function is represented by a table, similar to the way trigonometric functions and logarithms were represented for centuries. New differential equations to be solved keep arising in various branches of science, and they will lead to tables for new functions.

4. As approximated by interpolation.

The values given in tables of functions are usually only approximate. However, this is also true of the values given by a calculator. If the table is prepared with sufficient care, the approximate values listed can be as accurate as, or more than, those

4. As approximated by interpolation

Table 3. 1

x	e^x	e^{-x}	x	e^x	e^{-x}
0. 00	1. 0000	1. 0000	2. 5	12. 182	0. 0821
0. 05	1. 0513	0. 9512	2. 6	13. 464	0. 0743
0. 10	1. 1052	0. 9048	2. 7	14. 880	0. 0672
0. 15	1. 1618	0. 8607	2. 8	16. 445	0. 0608
0. 20	1. 2214	0. 8187	2. 9	18. 174	0. 0550
0. 25	1. 2840	0. 7788	3. 0	20. 086	0. 0498
0. 30	1. 3499	0. 7408	3. 1	22. 198	0. 0450
0. 35	1. 4191	0. 7047	3. 2	24. 533	0. 0408
0. 40	1. 4918	0. 6703	3. 3	27. 113	0. 0369
0. 45	1. 5683	0. 6376	3. 4	29. 964	0. 0334
0. 50	1. 6487	0. 6065	3. 5	33. 115	0. 0302
0. 55	1. 7333	0. 5769	3. 6	36. 598	0. 0273
0. 60	1. 8221	0. 5488	3. 7	40. 447	0. 0247
0. 65	1. 9155	0. 5220	3. 8	44. 701	0. 0224
0. 70	2. 0138	0. 4966	3. 9	49. 402	0. 0202
0. 75	2. 1170	0. 4724	4. 0	54. 598	0. 0183
0. 80	2. 2255	0. 4493	4. 1	60. 340	0. 0166
0. 85	2. 3396	0. 4274	4. 2	66. 686	0. 0150
0. 90	2. 4596	0. 4066	4. 3	73. 700	0. 0136
0. 95	2. 5857	0. 3867	4. 4	81. 451	0. 0123
1. 0	2. 7183	0. 3679	4. 5	90. 017	0. 0111
1. 1	3. 0042	0. 3329	4. 6	99. 484	0. 0101
1. 2	3. 3201	0. 3012	4. 7	109. 95	0. 0091
1. 3	3. 6693	0. 2725	4. 8	121. 51	0. 0082
1. 4	4. 0552	0. 2466	4. 9	134. 29	0. 0074
1. 5	4. 4817	0. 2231	5	148. 41	0. 0067
1. 6	4. 9530	0. 2019	6	403. 43	0. 0025
1. 7	5. 4739	0. 1827	7	1096. 6	0. 0009
1. 8	6. 0496	0. 1653	8	2981. 0	0. 0003
1. 9	6. 6859	0. 1496	9	8103. 1	0. 0001
2. 0	7. 3891	0. 1353	10	22026	0. 00005
2. 1	[illegible]	[illegible]			
2. 2	9. 0250	0. 1108			
2. 3	9. 9742	0. 1003			
2. 4	11. 023	0. 0907			

given by a calculator, though usually they are not. So a table is not necessarily inferior to a calculator on the score of accuracy of values. Usually the major drawback of a table is the relatively small number of values listed. Thus, in Table 3. 1, values of e^x are given for only 66 different values of x. Just by reading values from the table, there would be only 66 values of x for which one would have an approximation for e^x.

Most likely, a value for x for which one would wish an estimate of e^x would fail to be one of the 66 values appearing in the table.

Fortunately there is interpolation. By interpolation, one can find approximate function values for values of x lying between those given in the table. You should remember a particular way to interpolate from your trigonometry course. If not, a description of it is given in Sect. 3 of Chap. VI.

This sort of interpolation is called linear interpolation because it depends on connecting two points of the graph of the function by a straight line. By suitable use of the calculator, one can perform improved types of interpolation. This gives better approximations for the function values for values of x lying between those given in the table. Thus, the approximation of a function by a table is considerably improved.

There are functional relationships occurring in engineering, physics, and other branches of science for which no formula is known, nor any means of calculating an approximation. By measurement and experiment, some values of x and y will be determined, and listed in a table. Further measurements and experiments would be expensive, but we wish to know more about the functional relationship. Interpolation can be very useful in such a case.

Getting back to the temperature at Greenwich Observatory, if there should happen to be a record of the hourly temperatures for July 3, 4, and 5 in 1776, then a suitable interpolation procedure should give a fairly accurate estimate of the temperature at the signing of the Declaration of Independence. Possibly this calculation is not as intractable as was suggested earlier.

A discussion of interpolation in general is given in Chap. VI and Chap. XVII.

Problem 4.1. By linear interpolation in Table 3.1, get an estimate of $e^{0.6625}$. Compare with the value for $e^{0.6625}$ given by the calculator.

Hint (for those who have forgotten how to do linear interpolation). Since 0.6625 is one quarter of the way from the listed x-value 0.65 to the listed x-value 0.70, one would expect that $e^{0.6625}$ would be about one quarter of the way from the value of 1.9155 listed for $e^{0.65}$ to the value of 2.0138 listed for $e^{0.70}$.

Chapter II

EVALUATION OF A POLYNOMIAL

0. Guide for the reader.

By all odds, the functions that you are most often called upon to evaluate are polynomials. Special procedures for doing this efficiently will be given in Sect. 1. As soon as you encounter a place in the calculus course where you are asked to get the value of a polynomial more complicated than a quadratic, you should familiarize yourself with Sect. 1, and be prepared to use the procedures given there.

The topics of Sect. 2, synthetic division and deflation, come up in situations such as taking the limit of the quotient of two polynomials. When you first read Sect. 1, it would be advisable to skim through Sect. 2, to get the general idea. Then, when you encounter the places in calculus where you should use the procedures of Sect. 2, the formulas should look familiar, and should remind you to make a more careful study of Sect. 2.

Sect. 3 should be read after you are familiar with differentiating a polynomial. It is common in calculus courses to emphasize how information about the derivative of f can be used in plotting the graph $y = f(x)$. For example, this topic takes up about one third of Chap. 3 of T-F. When you reach this point in your calculus course, you should read Sect. 3 with some care, as it is devoted to calculator examples of just this subject.

1. Horner's method.

A polynomial, p, of degree n or less can always be defined by the formula

$$p(x) = a_0x^n + a_1x^{n-1} + \dots + a_{n-1}x + a_n \tag{1.1}$$

for certain numerical coefficients $a_0, a_1, \dots, a_n$. Note that we have here indexed the coefficients in such a way that the index of the coefficient plus the corresponding exponent of x is n.

We consider the evaluation of the polynomial defined by (1.1) at some point c, so as to get $p(c)$. This means that we must put c for x in (1. 1) and calculate

$$a_0c^n + a_1c^{n-1} + \dots + a_{n-1}c + a_n$$

This does not look like a very formidable calculation. With the y^x key, we can calculate $c^n, c^{n-1}, \dots, c^1, c^0$. We multiply them respectively by $a_0, a_1, \dots, a_{n-1}, a_n$, and add the products. But stop and think. It takes three or four key strokes for each use of the y^x key, and we also have n multiplications and n additions. Besides which, the y^x key is relatively slow in operation, and tends to be the least accurate key on many inexpensive machines.

There is a better way to evaluate p at c from (1. 1). The trick is to rearrange

$$p(c) = a_0c^n + a_1c^{n-1} + \dots + a_{n-1}c + a_n$$

in "nested form", as

$$p(c) = ((\dots((a_0c + a_1)c + a_2)c + \dots + a_{n-2})c + a_{n-1})c + a_n \,. \tag{1. 2}$$

Now there is no need for the exponentiation key, y^x. Since it turns out that we can arrange the calculation so as also to avoid using any parenthesis keys, all we need to do is to multiply and add n times each.

We start the calculation by evaluating the innermost parenthetical group, and then work our way from the inside out. This means that we first calculate

$$b_1 = a_0c + a_1 \,.$$

From it, we calculate

$$b_2 = (a_0c + a_1)c + a_2 = b_1c + a_2 \,.$$

From it, we calculate

$$b_3 = ((a_0c + a_1)c + a_2)c + a_3 = b_2c + a_3 \,,$$

and so on, until we reach

$$\begin{aligned} b_n &= ((\dots((a_0c + a_1)c + a_2)c + \dots + a_{n-2})c + a_{n-1})c + a_n \\ &= b_{n-1}c + a_n \,. \end{aligned}$$

Clearly, by (1. 2), $b_n = p(c)$, the value of p at c, which is what we wish.

For example, let us calculate the value of

1. Horner's method

(1. 3) $$p(x) = 3x^3 - 5x^2 - 4$$

at $x = 2$. Here $a_0 = 3$, $a_1 = -5$, $a_2 = 0$ and $a_3 = -4$. We write $p(x)$ in nested form as

$$p(x) = ((3x - 5)x + 0)x - 4 .$$

With $c = 2$, we then calculate

$$b_1 = a_0 c + a_1 = (3)(2) - 5 = 1$$

$$b_2 = b_1 c + a_2 = (1)(2) + 0 = 2$$

$$b_3 = b_2 c + a_3 = (2)(2) - 4 = 0 .$$

Thus $p(2) = 0$. Note how the missing $a_2 x^1$ term in (1. 3) does appear in the calculation, as addition of a zero coefficient when forming b_2.

If we put $x = 2$ in (1. 3), we get

$$\begin{aligned} p(2) &= (3)2^3 - (5)2^2 - 4 \\ &= (3)(8) - (5)(4) - 4 \\ &= 24 - 20 - 4 = 0 . \end{aligned}$$

This agrees with our value for b_3 (as it should).

We now analyze and formalize this evaluation procedure. For convenience, we define

$$b_0 = a_0 .$$

Then the calculation proceeds through the following $n + 1$ steps:

$$\begin{aligned} b_0 &= a_0 \\ b_1 &= b_0 c + a_1 \\ b_2 &= b_1 c + a_2 \\ b_3 &= b_2 c + a_3 \\ &\cdots\cdots \\ b_{n-2} &= b_{n-3} c + a_{n-2} \\ b_{n-1} &= b_{n-2} c + a_{n-1} \\ b_n &= b_{n-1} c + a_n . \end{aligned}$$

Every line after the first follows the uniform format:

(1. 4)
$$b_r = cb_{r-1} + a_r .$$

In view of this, the calculation proceeds very simply. Start by inputting b_0 (that is, a_0) into the display. Then repeat n times the two following operations:

multiply by c

add a_r $\quad (r = 1, 2, 3, \ldots, n)$.

As we said, n multiplications and n additions, plus inputting c and the various a_r's, takes care of the whole thing. At the end, we have b_n, which is $p(c)$.

It is customary to call a specified sequence of computational steps, such as the above, an algorithm. This particular algorithm is called Horner's method. It is much the fastest way to calculate $p(c)$.

When an algorithm of this sort proceeds by repeating a given set of steps over and over, using the answer for the previous step as input for the next, then the algorithm is called recursive. Horner's method is a good example of a recursive algorithm. On a programmable calculator with subroutine capability, one can often use this recursive feature to shorten the program for the algorithm. This is done in the implementations of Horner's method given as Programs II. 1 and II. 2 in the Program Appendix.

RPN

Horner's method works especially well on an RPN calculator. Suppose one has the stack filled as in Table 1. 1.

Table 1. 1

t	c
z	c
y	c
x	b_{r-1}

Then the program

(1. 5)
$$\boxed{\times},\ a_r,\ \boxed{+}$$

will replace the b_{r-1} in the stack by b_r. Then a repetition of Program (1. 5) with $r + 1$ in place of r, namely

$$\boxed{\times},\ a_{r+1},\ \boxed{+}\ ,$$

will in turn replace the b_r by b_{r+1}. And so on.

RPN

RPN

So we start by filling the stack with values of c, which is done by inputting c and pressing ↑ three times. Then input a_0 (which is b_0). We are now ready to get $b_1, b_2, b_3, \ldots, b_n$ successively in the display, by repeating the Program (1. 5) with $r = 1, 2, \ldots, n$. Thus, on a nonprogrammable calculator, after entering c in the display, the rest of the calculation follows Program (1. 6).

(1. 6)

[↑], [↑], [↑], a_0,

[×], a_1, [+],

[×], a_2, [+],

. . . .

[×], a_{n-1}, [+],

[×], a_n, [+].

At the end of the calculation, $b_n = p(c)$ is in the display.

RPN

AE

On an AE calculator, the basic step, (1. 4), of Horner's method may be carried out as follows. Suppose we have b_{r-1} in the display. Then the program

[×], c, [+], a_r, [=]

will bring b_r into the display instead. But this requires inputting c at every step. If c is an "easy" number, like 3, it can be input by pressing a single key, and one can do no better. However, c could be a 10-digit number, so that there is risk of error every time one inputs it. Even on a nonprogrammable calculator, there is usually at least one memory register. So let us assume that we can store c. Then we calculate b_r from b_{r-1} by the program

(1. 5)

[×], [RCL] c, [+], a_r, [=].

Here we have written [RCL] c to indicate that the value of c is to be recalled into the display from wherever it has been stored. Thus, on a nonprogrammable calculator, after entering c in the display, the rest of the calculation follows Program (1. 6).

(1. 6)

[STO] c, a_0,

[×], [RCL] c, [+], a_1, [=],

[×], [RCL] c, [+], a_2, [=],

. . . .

[×], [RCL] c, [+], a_{n-1}, [=],

[×], [RCL] c, [+], a_n, [=].

At the end of the calculation, $b_n = p(c)$ is in the display.

AE

II. EVALUATION OF A POLYNOMIAL

Suppose now that you have been given the coefficients and wish to evaluate the polynomial defined by (1. 1) for several different values of c. On a nonprogrammable calculator, you simply repeat (1. 6) for each different value of c. This requires inputting the whole succession of a_r's for each value of c. If the a_r's are 10-digit numbers, which they may well be, this is quite a chore, and entails considerable risk of error with all that inputting.

If you have a programmable calculator, you can arrange to store Program (1. 6). Then, upon pressing the program key (or keys), the calculator will run through Program (1. 6) for you, including inputting the various a_r's. (For more details, refer to Chap. 0.) So now, if you wish to calculate $p(c)$ for some c's, you input any given c into the display, press the program key (or keys), and sit back and wait for b_n, which equals $p(c)$, to appear in the display.

In Chap. 0, you are advised to check the program after recording it. Let us again stress this. For polynomials, there is an additional test that you can use, and which you should use. It is easy to calculate $p(1)$, since it is just the sum of the coefficients. So, after you have stored the program, and have checked it, run yet another test by taking $c = 1$ and seeing if b_n comes out equal to the sum of the coefficients.

If you have a high degree polynomial, there may be so many a_r's that Program (1. 6) will take more steps than your calculator has room to store. In that case, go as far as you can through Program (1. 6) with a stored program, arranging to stop with some b_r in the display. Then you can finish Program (1. 6) by hand, inputting a_r's as you go.

Nowadays, most programmable calculators have a subroutine capability. Because of the recursive nature of Horner's algorithm, this can be used to reduce the number of program registers required to store the program. Program II. 1 in the Program Appendix shows how to take advantage of this. You should use this program (or one similar to it that you have worked up for your own calculator) for the next problem.

Problem 1. 1. (Only for those with good programmable calculators.) Evaluate $p(c)$ for the polynomial, p, defined by

$$(1.7) \quad p(x) = x^{18} + 2x^{17} + 3x^{16} + 4x^{15} + 5x^{14} + 6x^{13} + 7x^{12} + 8x^{11} + 9x^{10} + 10x^{9} + 9x^{8} + 8x^{7} + 7x^{6} + 6x^{5} + 5x^{4} + 4x^{3} + 3x^{2} + 2x + 1$$

for $c = -3, -2, -1, 0, 1, 2, 3$.

Program (1. 6), or Program II. 1 of the Program Appendix, if used with a programmable calculator, involves storing the digits of the various a_r's as part of the program. For typical examples in calculus texts, each coefficient has only one or two digits, and so one can deal with polynomials of fairly high degree. However, one can have coefficients that use up as many as 15 program registers apiece (see Chap. 0). In such cases, it is advisable to store the coefficients $a_0, a_1, \ldots, a_n$ beforehand in memory registers, and in Program (1. 6) or Program II. 1 of the Program Appendix put [RCL] a_r where now a_r is indicated to be worked into the program. In doing this, one is limited by the number of memory registers available. This is embodied in Program II. 2 of the Program Appendix.

Problem 1. 2. In Prob. 11 at the end of Sect. 1 - 9 of T- F, there is given a table of values of a function:

s (in ft)	10	38	58	70	74	70	58	38	10
t (in sec)	0	0.5	1.0	1.5	2.0	2.5	3.0	3.5	4.0

Show by calculating its values that the polynomial, s, defined by

$$s(t) = -16t^2 + 64t + 10 \tag{1.8}$$

agrees with the s of the table for the values of t listed. Remember to use Horner's method to evaluate the polynomial. That is, for the given values of t, take

$$s(t) = (-16t + 64)t + 10 .$$

The fact that (1. 8) holds for the specified values of t gives no assurance whatever that it holds for other values of t. However, it is perhaps a not unreasonable assumption that (1. 8) was intended to hold for general values of t.

Problem 1. 3. For the polynomial, p, defined by

$$p(x) = x^4 - 4x^3 + 2x^2 - 4x + 1 , \tag{1.9}$$

calculate $p(c)$ for $c = -2, -1, 0, 1, 2, 3, 4, 5, 6$. If you do not own a programmable calculator, do it for $c = 0, 1, 2, 3, 4$. (Calculate $p(0)$ in your head, of course.) If you do own a programmable calculator, enlarge the program so that it will also generate the values of c for you, with convenient stops and restarts, so that you do not have to input nine different values of c. Make a rough sketch of the graph of $y = p(x)$. Save this graph for subsequent use.

Remark. There is much stress in calculus on getting a careful determination of the graph of $y = f(x)$, by observing when the derivative of f is positive, negative, or zero. The polynomial, p, defined by (1. 9) is the derivative of the polynomial, P, defined by

$$P(x) = \frac{x^5}{5} - x^4 + \frac{2x^3}{3} - 2x^2 + x - 2 . \tag{1.10}$$

(If you cannot presently verify this, you will very soon be taught how.) To get a careful determination of the graph of $y = P(x)$, you will need to know when the derivative of P, namely p, has values which are positive, negative or zero.

Problem 1. 1. For the polynomial, p, defined by (1. 9), use your graph of it to guess very approximately the values of c (if any) for which $p(c) = 0$, and to identify the values of x for which $p(x)$ is positive, and the values of x for which $p(x)$ is negative.

Remark. The question of determining more accurately the values of c for which $p(c) = 0$ is taken up in Chap. VII. For the present, guess them as well as you can by eyeballing the graph. If $p(c) = 0$, we say that c is a zero of p.

Problem 1.5. For the polynomial, p, defined by

$$p(x) = 8x^4 - 14x^3 - 9x^2 + 11x - 1 \tag{1.11}$$

calculate $p(c)$ for $c = -2, -1, 0, 1, 2, 3$. Make a rough sketch of the graph of $y = p(x)$. Save this graph for subsequent use. Use your graph to guess very approximately the values of c (if any) for which $p(c) = 0$, and to identify the values of x for which $p(x)$ is positive, and the values of x for which $p(x)$ is negative.

Problem 1.6. Even if you do not have a programmable calculator, calculate $p(c)$ for $c = 0, 1,$ and 2 for the p of Prob. 1.1.

Hint. For that p, verify that

$$p(x) = \{x^9 + x^8 + x^7 + x^6 + x^5 + x^4 + x^3 + x^2 + x + 1\}^2 . \tag{1.12}$$

Remark. We cannot stress too strongly that the reader should remain alert to see where some trick, such as above, can cut the labor of computation quite a bit.

Problem 1.7. Even if you do not have a programmable calculator, calculate $p(c)$ for $c = -3, -2, -1, 0, 1, 2,$ and 3 for the p of Prob. 1.1.

Remark. We just barely got through saying that you should remain alert to see where a trick can cut the labor of computation. Verify (by multiplying both sides by $x - 1$) that if $x \neq 1$, then

$$\frac{x^{10} - 1}{x - 1} = x^9 + x^8 + x^7 + x^6 + x^6 + x^4 + x^3 + x^2 + x + 1 . \tag{1.13}$$

2. Synthetic division and deflation.

As we will show in a moment, Horner's method provides an efficient way to divide the right hand side of (1.1) by the linear expression $x - c$, if possible. There are several reasons why one might want to carry out such a division.

For example, in trying to determine

$$\lim_{x \to c} p(x)/q(x)$$

with p and q both polynomials, one may have to deal with the fact that both p and q are zero at the point c. Then one must compute polynomials p_1 and q_1 so that $p_1(x)(x - c) = p(x)$ and $q_1(x)(x - c) = q(x)$ and consider

$$\lim_{x \to c} p_1(x)/q_1(x) .$$

As another example, consider the problem of finding all the zeros of a polynomial p. One usually finds such zeros one at a time, for example by the methods discussed in

2. Synthetic division and deflation

Chap. VII. Having found a first zero, say, c_1, one can find the remaining zeros by finding the zeros of the "reduced" polynomial p_1, defined by $p_1(x) = p(x)/(x-c_1)$. As soon as one has found a zero of p_1, say c_2, one goes over to a further reduced polynomial p_2, defined by $p_2(x) = p_1(x)/(x-c_2)$, etc. In this way, the polynomials whose zeros are to be found become ever simpler. In the end, one has a factorization $p(x) = a_0(x-c_1)(x-c_2)\ldots(x-c_n)$. Such a factorization is needed further on in the calculus, when integrating rational functions by the method of partial fraction expansions.

The connection between Horner's method and the division of $p(x)$ by a linear expression $x - c$ is contained in the following theorem.

Theorem 2.1. Let $b_0, b_1, \ldots, b_n$ be the quantities calculated during Horner's algorithm for evaluating

(2.1) $$p(x) = a_0x^n + a_1x^{n-1} + \ldots + a_{n-1}x + a_n$$

at $x = c$. This means that

(2.2a) $$b_0 = a_0$$

and

(2.2b) $$b_r = b_{r-1}c + a_r, \qquad \text{for } r = 1,\ldots,n.$$

Then

(2.3) $$p(x) = q(x)(x - c) + b_n$$

with q the polynomial defined by

(2.4) $$q(x) = b_0x^{n-1} + b_1x^{n-2} + \ldots + b_{n-2}x + b_{n-1}.$$

To see the truth of this theorem, multiply out $(x-c)q(x) + b_n$. By appealing to (2.4), this gives:

$$\begin{aligned} xq(x) &= xb_0x^{n-1} + xb_1x^{n-2} + xb_2x^{n-3} + \ldots + xb_{n-1} \\ -cq(x) &= \qquad\qquad - cb_0x^{n-1} - cb_1x^{n-2} - \ldots - cb_{n-2}x - cb_{n-1} \\ +b_n &= \qquad\qquad\qquad\qquad\qquad\qquad\qquad\qquad + b_n . \end{aligned}$$

On the right, combine each term with the one below (or above) it, to get:

$$b_0x^n + (b_1 - cb_0)x^{n-1} + (b_2 - cb_1)x^{n-2} + \ldots + (b_{n-1} - cb_{n-2})x + (b_n - cb_{n-1}) .$$

But $b_0 = a_0$ by (2.2a), and by (2.2b)

$$\begin{aligned} b_1 - cb_0 &= a_1 \\ b_2 - cb_1 &= a_2 \\ \cdots & \\ b_{n-1} - cb_{n-2} &= a_{n-1} \\ b_n - cb_{n-1} &= a_n \end{aligned} .$$

So we get

$$(x - c)q(x) + b_n$$
$$= a_0x^n + a_1x^{n-1} + \ldots + a_{n-1}x + a_n ,$$

as claimed in (2. 3).

<u>Corollary 1.</u> If p is a polynomial, and $p(c) = 0$, then $x - c$ divides $p(x)$ exactly.

<u>Proof.</u> If $p(c) = 0$, that means that $b_n = 0$. Putting $b_n = 0$ in (2. 3) gives

(2. 5) $$p(x) = (x - c)q(x) ,$$

which is just what the corollary claims to be the case.

<u>Corollary 2.</u> If p is a polynomial, $c_1 \neq c_2$, and $p(c_1) = p(c_2) = 0$, then the quadratic expression $(x - c_1)(x - c_2)$ divides $p(x)$ exactly.

<u>Proof.</u> Using Cor. 1, we can take $c = c_1$ in (2. 5). Then put $x = c_2$ in (2. 5):

$$p(c_2) = (c_2 - c_1)q(c_2).$$

But $p(c_2) = 0$ and $c_2 - c_1 \neq 0$, so that $q(c_2) = 0$. So, by Cor. 1, there is a polynomial r such that

$$q(x) = (x - c_2)r(x).$$

If we substitute this into (2. 5), with $c = c_1$, we get

(2. 6) $$p(x) = (x - c_1)(x - c_2)r(x) ,$$

which is just what the corollary claims to be the case.

Obviously, if we are going to make use of Thm. 2. 1, we have to make a record of the values of the b_r's. If we are not calculating the b_r's by a stored program, but are going along inputting the a_r's at every step, there is plenty of time to copy off the b_r's as they appear in the display. However, suppose we have in mind using Program II. 1 or II. 2 of the Program Appendix, or something analogous for another calculator. It is usually considered an advantage of Programs II. 1 and II. 2 that they go rapidly through the calculation, and all that one sees is b_n, which is $p(c)$. But now we would like to see, and copy down, the various b_r's. A very simple modification of Programs II. 1 or II. 2 will permit this. Just add [R/S] appropriately in the subroutine, as explained in the Program Appendix. Now the program will stop every time a b_r appears in the display. After copying it down, press R/S , and the program will go on to the next b_r. Check your copying.

If there is a copious supply of memory locations, so that the a_r's fill up no more than about half the registers, one could even arrange to have the b_r's stored in some of the vacant registers. This would be particularly useful if one has in mind further calculations involving the b_r's. It is easy enough to modify Program (1. 6) to do this by inserting suitable storage commands at the strategic points. Programs II. 1 and II. 2 would require extensive alterations, since the storage of the b_r's would interrupt the subroutine differently each time the subroutine is used. However, with the program set to stop at each b_r, one could store them one by one by hand in unused registers during the stops.

If one is not going to use the a_r's again, the b_r's could be put in their place. Details are given in Program II. 3 of the Program Appendix.

Illustration. Let us find

(2. 7)
$$\lim_{x \to -2} \frac{x^2 + x - 2}{x^2 + 5x + 6} .$$

As x approaches -2, both the numerator and denominator approach 0, so that it looks as though we are stuck with trying to guess a value for

$$\frac{0}{0} .$$

But Cor. 1 for Thm. 2. 1 says that if a polynomial, p, is 0 at the point -2, then $p(x)$ must be exactly divisible by $x + 2$. One can work out what the other factor is by the algorithm of the previous section, as explained in Thm. 2. 1, but for quadratics, such as appear in (2. 7), the other factor is obvious. So we rewrite (2. 7) as

$$\lim_{x \to -2} \frac{(x+2)(x-1)}{(x+2)(x+3)} = \lim_{x \to -2} \frac{x-1}{x+3}$$

$$= \frac{-2 - 1}{-2 + 3} = -3 .$$

Problem 2. 1. Evaluate:

(a)
$$\lim_{x \to -3} \frac{x^2 + 4x + 3}{x + 3}$$

(b) $\lim_{x \to 2} \frac{x^2 + 3x - 10}{x - 2}$

(c) $\lim_{x \to 1} \frac{x - 1}{2x^2 - 7x + 5}$

(d) $\lim_{x \to 2} \frac{x^3 - 8}{x^2 - 4}$

(e) $\lim_{x \to -1} \frac{2x^2 + 7x + 5}{x^3 + 1}$

Problem 2.2. For $n = 3, 4, 5, 6$ evaluate

$$\lim_{x \to 1} \frac{x^n - 1}{x^2 - 1}$$

Before you get up to $n = 6$ you should be able to figure out how to calculate

$$\lim_{x \to 1} \frac{x^n - 1}{x^2 - 1}$$

for general n; as soon as you do, carry out the calculation for general n, and get on to the next problem.

Problem 2.3. For the polynomial, p, defined by

(2.8) $$p(x) = x^4 - 4x^3 + 2x - 4x + 1$$

and for $c = 4$, calculate q and b_n such that (2.3) holds.

Problem 2.4. For the polynomial, p, defined by

(2.9) $$p(x) = 8x^4 - 14x^3 - 9x^2 + 11x - 1$$

calculate q_i and d_i such that

(2.10) $$p(x) = (x+1)q_1(x) + d_1$$

(2.11) $$p(x) = (x-3)q_2(x) + d_2 .$$

In Prob. 1.4, you were asked to state from looking at the graph where $p(x)$ is positive or negative for the p defined by (1.9). However, the graph went only from $x = -2$ to $x = 6$ (or from $x = 0$ to $x = 4$ if you do not own a programmable calculator). So you had to guess outside of that range. However, one can easily establish the facts. Note that the polynomial defined by (1.9) is the same as the one defined by (2.8).

*Problem 2. 5. For the polynomial, p, defined by (1. 9) and by (2. 8), show that $p(x)$ is positive for $x < 0$ and for $x > 4$.

Hint. If x is negative, can any term of $p(x)$ be negative? If $x > 4$, note that the $q(x)$ and b_n of Prob. 2. 3 are both positive.

*Problem 2. 6. For the polynomial, p, defined by (1. 11) and by (2. 9), show that $p(x)$ is positive for $x < -1$ and for $x > 3$.

Hint. If $x < -1$, note that the d_1 of (2. 10) is positive, while the $q_1(x)$ is negative. If $x > 3$, note that the d_2 and $q_2(x)$ of (2. 11) are both positive.

Consider the polynomial, p, defined by

$$p(x) = 12x^2 - 24x + 4 . \tag{2. 12}$$

According to the quadratic formula, which is (IV. 3. 1), $p(c_i) = 0$ for

$$c_1 = \frac{3+\sqrt{6}}{3}, \qquad c_2 = \frac{3-\sqrt{6}}{3} . \tag{2. 13}$$

So, by Cor. 2 to Thm. 2. 1, $(x - c_1)(x - c_2)$ should divide $p(x)$ exactly. It is easy to see, by multiplying out, that indeed

$$p(x) = 12(x - \frac{3+\sqrt{6}}{3})(x - \frac{3-\sqrt{6}}{3}) . \tag{2. 14}$$

Problem 2. 7. Using your calculator, calculate

$$c_1 = \frac{3+\sqrt{6}}{3}, \qquad c_2 = \frac{3-\sqrt{6}}{3} .$$

Suppose with these approximations for c_1 and c_2, your calculator seems to give exactly, by calculation,

$$p(x) = 12(x-c_1)(x-c_2). \tag{2. 15}$$

Would that prove that (2. 14) is true? As a matter of fact, does your calculator give (2. 15) EXACTLY, with the approximations you calculated for c_1 and c_2? Use all the digits that your calculator carries, and not just what it displays.

Problem 2. 8. Show that

$$96x^2 - 84x - 18 = 96(x - \frac{7+\sqrt{97}}{16})(x - \frac{7-\sqrt{97}}{16}) . \tag{2. 16}$$

Sometimes we wish to "deflate" a polynomial. This will happen after we have found that c is a zero of the polynomial p, so that $p(c) = 0$. Then, by Cor. 1 to Thm. 2. 1, we have

$$p(x) = (x - c)q(x) .$$

If we have the coefficients of $p(x)$ stored in the memory registers, we may wish to discard them, and put the coefficients of $q(x)$ in their place. This would happen if we are using Program II. 2. In the Program Appendix, we give Program II. 3 which will substitute the coefficients of $q(x)$ for those of $p(x)$ in the memory registers.

If we are using Program (1. 6) with a nonprogrammable calculator, the question would not arise. We start with the coefficients of $p(x)$ written on a piece of paper. As we work our way through Program (1. 6), we generate the coefficients of $q(x)$, which we write on a new piece of paper.

Problem 2. 9. With the polynomial, p, defined by

$$p(x) = 8x^3 + 2x^2 - 5x + 1 , \tag{2. 17}$$

we have $p(-1) = 0$. So there is a q such that

$$p(x) = (x + 1)q(x) .$$

Start with the coefficients of $p(x)$ stored in the calculator and run a program to get the coefficients of $q(x)$ stored there instead. (If you have a nonprogrammable calculator, simply list the coefficients of $q(x)$.)

3. Derivative of a polynomial.

The formula (2. 3), namely

$$p(x) = (x - c)q(x) + b_n ,$$

will be of use when we wish to evaluate both $p(x)$ and $p'(x)$ at the same point, $x = c$, where p' is the derivative of p. If we differentiate both sides of the equation above with respect to x, and use the formula for the derivative of a product (see Rule 5 of Sect. 2 - 3 of T - F), we will get

$$p'(x) = q(x) + (x - c) q'(x) ,$$

since the derivative of the constant b_n is zero. Taking $x = c$ gives

$$p'(c) = q(c) . \tag{3. 1}$$

This means that we can compute $p'(c)$ by evaluating the polynomial $q(x)$ at c (by Horner's method, of course).

The above procedure gives the most efficient way to evaluate both a polynomial and its first derivative at the same point c. If we need to evaluate only the first derivative, then it is more efficient to form the polynomial p' directly and evaluate it by Horner's method.

3. Derivative of a polynomial

Problem 3. 1. For the q and p of Prob. 2. 3, calculate $q(4)$ and $p'(4)$, and see if they agree.

Problem 3. 2. For the q_i and p of Prob. 2. 4, see if $q_1(-1)$ agrees with $p'(-1)$, and if $q_2(3)$ agrees with $p'(3)$.

When we undertake to get the value of the derivative of p at the point c by evaluating $q(c)$, that requires knowledge of the b_r's. So at first thought, it seems as if we have to do a lot of storing of coefficients, namely all the b_r's. However, if we manage things efficiently, this is not so.

Recalling what q is from Thm. 2. 1, we see that Horner's method for calculating $q(c)$ involves the following algorithm:

$$\begin{aligned} c_0 &= b_0 \\ c_1 &= cc_0 + b_1 \\ c_2 &= cc_1 + b_2 \\ &\cdots\cdots \\ c_{n-1} &= cc_{n-2} + b_{n-1} . \end{aligned}$$

Then $q(c) = c_{n-1}$.

The trick is to calculate the b_r's and c_r's in parallel. Really, it is b_{r+1} and c_r that go in parallel. Then two registers will suffice for their storage. Suppose b_{r+1} is in one register and c_r is in the other. One uses b_{r+1} and c_r to calculate c_{r+1}, which one stores in place of c_r. Then one uses b_{r+1} and a_{r+2} to calculate b_{r+2}, which one stores in place of b_{r+1}. Then one starts over again.

At the beginning, one does a little calculation to get b_1 and c_0, and to store them. Then one iterates the process above until one gets up to b_n and c_{n-1}, which are $p(c)$ and $p'(c)$.

Presumably, one has c stored someplace, so that one does not need to input it twice for each iteration.

RPN

On an RPN calculator, one can manage to keep c stored in the stack, so that two memory registers will suffice, one to hold the b_r's and one to hold the c_r's. Then the following fairly simple program will suffice

c, [↑], [↑], [↑],

a_0, [STO] c_0, [×], a_1, [+], [STO] b_1,

[CLX], [RCL] c_0, [×], [RCL] b_1, [+], [STO] c_1,

[CLX], [RCL] b_1, [×], a_2, [+], [STO] b_2,

[CLX], [RCL] c_1, [×], [RCL] b_2, [+], [STO] c_2,

RPN

RPN

[CLX], [RCL] b_2, [×], a_3, [+], [STO] b_3,

. . . .

[CLX], [RCL] c_{n-2}, [×], [RCL] b_{n-1}, [+], [STO] c_{n-1},

[CLX], [RCL] b_{n-1}, [×], a_n, [+], [STO] b_n .

The CLX that appears is the command to clear the display. The first line of the program fills the stack with c's. The second key stroke of the second line stores a_0 in the register reserved for the c_r's. But $a_0 = c_0$, so that then we have c_0 appropriately stored. After the [+] of the second line, we have b_1 in the display, and it is proper then to store it in the register reserved for the b_r's.

After the [+] in the third line, we have c_1 in the display. The final command, [STO] c_1, stores it in the register reserved for the c_r's. This obliterates c_0, which is fine because we do not need c_0 any more. The fourth line calculates b_2 and stores it in place of b_1. This is also fine, as we do not need b_1 any more.

The fifth line calculates c_2 and puts it in place of c_1, while the sixth line calculates b_3 and puts it in place of b_2. Right on!

So we go smoothly along, and at the end we have c_{n-1}, which is $p'(c)$, and b_n, which is $p(c)$.

You reserve whatever registers happen to be convenient to store the b_r's and c_r's. If you had happened to pick registers one and two respectively, then the program shown above can be put into the schematic form embodied in Program (3. 2).

(3. 2) Preparation:

c, [↑], [↑], [↑],

a_0, [STO 2], [×], a_1, [+], [STO 1],

Loop, to be repeated for $r = 0, 1, \ldots, n-2$:

[CLX], [RCL 2], [×], [RCL 1], [+], [STO 2],

[CLX], [RCL 1], [×], a_{r+2}, [+], [STO 1] .

If you have b_{r+1} in register one and c_r in register two, then the first line of the Loop puts c_{r+1} into register two and the second line puts b_{r+2} into register one. You repeat the Loop $n-1$ times, with $r = 0, 1, \ldots, n-2$. Then you will have b_n, which is $p(c)$, in register one and c_{n-1}, which is $p'(c)$, in register two.

It should occur to you that there is a subprogram in this which is repeated many times, namely Program (3. 3).

(3. 3) [+], [STO 1], [CLX], [RCL 2], [×], [RCL 1]

[+], [STO 2], [CLX], [RCL 1], [×] .

RPN

3. Derivative of a polynomial

RPN Executions of this are interspersed with inputs (or recalls) of the a_r's. There is a Preparation at the beginning and a short Termination at the end. If you have a programmable calculator, this is the sort of thing you would store as a recallable program. On the more sophisticated calculators, it could be a subroutine.

By carrying out some stack gymnastics, you can carry all three of c, c_r, and b_{r+1} in the stack simultaneously, juggling these three quantities appropriately by using the R↓ and x ⇄ y keys. This avoids using up two registers for c_r and b_{r+1}, leaving more room to store the a_r's. If you like puzzles, then you'll enjoy figuring out key sequences to do such things. Actually, a great many such key sequences are given in the Table 2. 5. 1 on pp. 62-86 of "Algorithms for RPN calculators," by John A. Ball, John Wiley and Sons, 1978.

For the particular case at hand, you can go from the stack configuration shown in Table 3. 1 to that shown in Table 3. 2 by the program in the first line of Table 3. 3.

Table 3. 1

t	c
z	c_r
y	cb_r
x	a_{r+1}

Table 3. 2

t	c
z	c
y	c_{r+1}
x	cb_{r+1}

Table 3. 3

	+ ,	R↓ ,	× ,	x ⇄ y ,	R↓ ,	+ ,	R↓ ,	x ⇄ y ,	R↓ ,	×
c	c	b_{r+1}	b_{r+1}	b_{r+1}	c	c	c_{r+1}	c_{r+1}	c	c
c_r	c	c	b_{r+1}	b_{r+1}	b_{r+1}	c	c	c	c_{r+1}	c
cb_r	c_r	c	c	cc_r	b_{r+1}	b_{r+1}	c	b_{r+1}	c	c_{r+1}
a_{r+1}	b_{r+1}	c_r	cc_r	c	cc_r	c_{r+1}	b_{r+1}	c	b_{r+1}	cb_{r+1}

Below each indicated keystroke is a vertical line, on the left of which is what was in the stack before the stroke and on the right of which is what will be in the stack after the stroke.

Having got to the configuration of Table 3. 2, a subsequent input of a_{r+2} will bring the configuration of Table 3. 2 to agree with that of Table 3. 1, except for having r replaced by $r + 1$. So the program shown above can be used as a subprogram, to be used alternately with inputting (or recalling) values of a_s, for $s = 2, 3, \ldots, n$. After inputting (or recalling) a_n, execute [+], and you will have $p'(c)$ and $p(c)$ in the y-register and x-register respectively.
RPN

RPN

An actual program embodying these ideas is given as Program II. 4 in the Program Appendix.

RPN

AE

On an AE calculator, it would seem that you will need three memory registers, one to hold the b_r's, one to hold the c_r's. and one to hold c. If these are available, then the following fairly simple program will suffice.

c, [STO] c,

[×], a_0, [STO] c_0, [+], a_1, [=], [STO] b_1,

[RCL] c_0, [×], [RCL] c, [+], [RCL] b_1, [=], [STO] c_1,

[RCL] b_1, [×], [RCL] c, [+], a_2, [=], [STO] b_2,

[RCL] c_1, [×], [RCL] c, [+], [RCL] b_2, [=], [STO] c_2,

[RCL] b_2, [×], [RCL] c, [+], a_3, [=], [STO] b_3,

. . . .

[RCL] c_{n-2}, [×], [RCL] c, [+], [RCL] b_{n-1}, [=], [STO] c_{n-1},

[RCL] b_{n-1}, [×], [RCL] c, [+], a_n, [=], [STO] b_n .

The first line of the program stores c. The third key stroke of the second line stores a_0 in the register reserved for the c_r's. But $a_0 = c_0$, so that then we have c_0 appropriately stored. After the [=] of the second line, we have b_1 in the display, and it is proper then to store it in the register reserved for the b_r's.

After the [=] in the third line, we have c_1 in the display. The final command, [STO] c_1, stores it in the register reserved for the c_r's. This obliterates c_0, which is fine because we do not need c_0 any more. The fourth line calculates b_2 and stores it in place of b_1. This is also fine, as we do not need b_1 any more.

The fifth line calculates c_2 and puts it in place of c_1, while the sixth line calculates b_3 and puts it in place of b_2. Right on!

So we go smoothly along, and at the end we have c_{n-1}, which is $p'(c)$, and b_n, which is $p(c)$.

You reserve whatever registers happen to be convenient to store the b_r's, c_r's, and c. If you had happened to pick registers one, two, and three respectively, then the program shown above can be put into the schematic form embodied in Program (3. 2).

(3. 2) Preparation:

c, [STO 3] ,

[×], a_0, [STO 2], [+], [a_1], [=], [STO 1] ,

AE

AE

Loop, to be repeated for $r = 0, 1, \ldots, n-2$:

[RCL 2], [×], [RCL 3], [+], [RCL 1], [=], [STO 2],

[RCL 1], [×], [RCL 3], [+], a_{r+2}, [=], [STO 1].

If you have b_{r+1} in register one and c_r in register two, then the first line of the Loop puts c_{r+1} into register two and the second line puts b_{r+2} into register one. You repeat the Loop $n-1$ times, with $r = 0, 1, \ldots, n-2$. Then you will have b_n, which is $p(c)$, in register one and c_{n-1}, which is $p'(c)$, in register two.

It should occur to you that there is a subprogram in this which is repeated many times, namely Program (3. 3).

(3. 3) [=], [STO 1], [RCL 2], [×], [RCL 3], [+], [RCL 1],

[=], [STO 2], [RCL 1], [×], [RCL 3], [+].

Executions of this are interspersed with inputs (or recalls) of the a_r's. There is a Preparation at the beginning and a short Termination at the end. If you have a programmable calculator, this is the sort of thing you would store as a recallable program. On the more sophisticated calculators, it could be a subroutine.

On some calculators there is a register exchange key which enables you to carry along all three of b_{r+1}, c_r, and c without having to use more than two memory registers. This leaves one more register to store the a_r's. An actual program of this sort is given as Program II. 4 in the Program Appendix.

AE

As noted earlier, you could always calculate the coefficients of p'. Then you can calculate both $p(c)$ and $p'(c)$ by applying Horner's method to p and p' respectively. If you need both of $p(c)$ and $p'(c)$, the method given above is preferable for the following reason. To apply Horner's method for both p and p', you must input (or store) coefficients for both p and p', which are nearly twice as many as for p alone. With the method above, only the coefficients of p are involved.

In the calculus, textbooks recommend that, when one is graphing $y = f(x)$, one not only should plot some points (x_i, y_i), where $y_i = f(x_i)$, but that one should draw an arrow of slope $f'(x_i)$ through the point (x_i, y_i), to indicate the direction in which the curve is moving when it passes that point. Let us try this for the polynomial, p, defined by

$$p(x) = 8x^4 - 14x^3 - 9x^2 + 11x - 1 \quad (3.4)$$

using Program (3. 2) or Program II. 4. This is the polynomial that is defined in (1. 11).

In Prob. 1. 5, you were asked to draw a graph of $y = p(x)$ for this polynomial. You might naturally have produced something that looked a bit like Fig. 3. 1.

However, let us follow the suggestion of the calculus text, and use the slopes at

Figure 3. 1

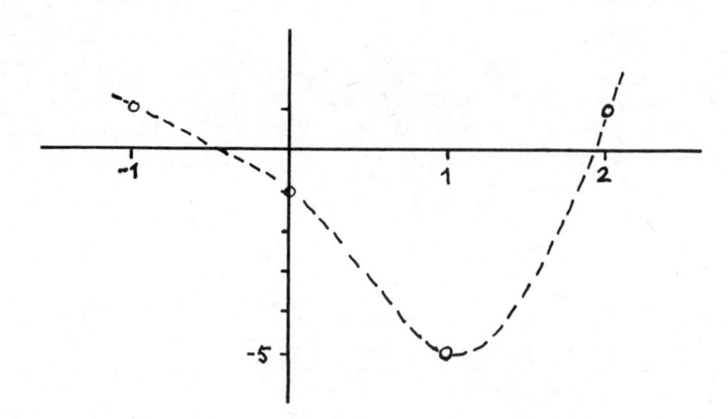

the four points shown. Values of $p(x)$ and $p'(x)$ are given in Table 3. 4. The disparity between the slopes from Table 3. 4 and those shown in Fig. 3. 1 is scandalous. With a

Table 3. 4

x	p(x)	p'(x)
-1	1	-45
0	-1	11
1	-5	-17
2	1	63

slope of +11 at the point (0, -1), it would seem as if the curve would have to get above the x-axis before x gets much larger. Let us try $x = 0.5$. We have $p(0.5) = 1$ and $p'(0.5) = -4.5$.

Let's face it. The curve shown in Fig. 3. 1 has an entirely wrong shape. So we had better give it the full treatment, as expounded in the calculus text. For this, we use Program (3. 2) or Program II. 4 and tabulate our polynomial given by (3. 4) and its derivative more finely between -1 and 2, say at the quarter points -1, -0. 75, -0. 5, -0. 25, . . . , 1. 5, 1. 75, 2. This gives Table 3. 5. From this table, it appears that p' vanishes somewhere in the intervals (-0. 75, -0. 5), (0. 25, 0. 5), and (1. 5, 1. 75). Since p' is a polynomial of degree 3, it cannot vanish anywhere else, by a well known theorem that says that a polynomial of degree n cannot vanish at more than n distinct points. We therefore know now approximately where $p'(x)$ is positive and where it is negative, that is, where p is increasing and where p is decreasing. Figure 3. 2

3. Derivative of a polynomial

Figure 3. 2

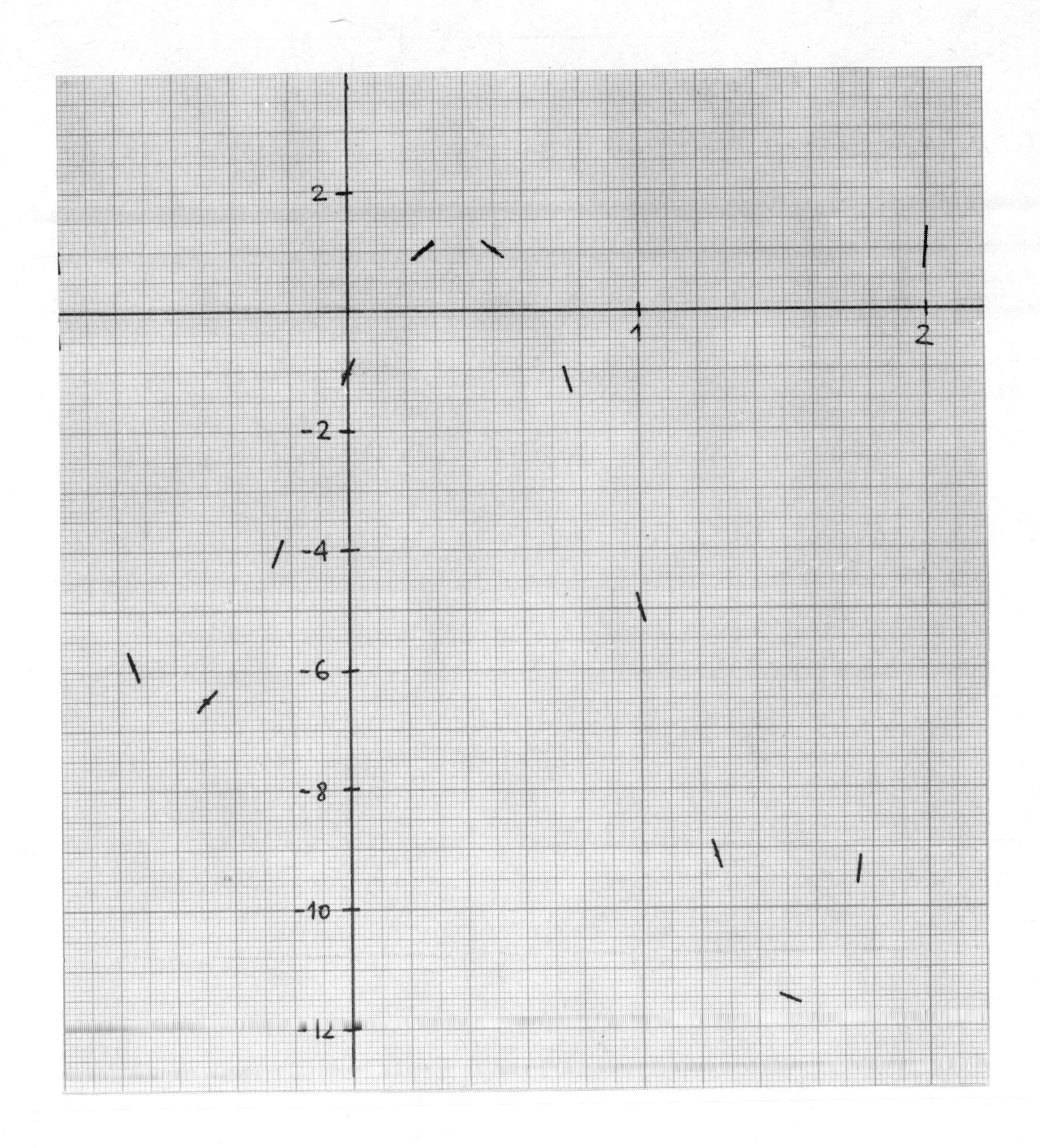

Table 3. 5

x	p(x)	p'(x)
-1	1	-45
-0. 75	-5. 875	-12. 625
-0. 5	-6. 5	5. 5
-0. 25	-4. 0625	12. 375
0	-1	11
0. 25	1	4. 375
0. 5	1	- 4. 5
0. 75	-1. 1875	-12. 625
1	-5	-17
1. 25	-9. 125	-14. 625
1. 5	-11. 5	- 2. 5
1. 75	-9. 3125	22. 375
2	1	63

contains the information from Table 3. 5 in graphical form, from which the shape of the graph becomes quite evident.

Problem 3. 3. For the polynomial, p, defined by (1. 9), namely by

$$p(x) = x^4 - 4x^3 + 2x^2 - 4x + 1,$$

calculate $p'(c)$ for $c = 0, 1, 2, 3, 4$. See if these values seem to agree fairly well with the slopes of the approximate graph of $y = p(x)$ that you drew for Prob. 1. 3.

Chapter III

DIFFERENCE QUOTIENTS

0. Guide for the reader.

Difference quotients are used to calculate slopes, and the derivative is defined as the limit of a certain difference quotient. Hence this short chapter is required reading as soon as either a slope or a derivative is encountered in the calculus course.

The most commonly occurring difference quotient is the one used to define the derivative, namely

$$\frac{f(x + \Delta x) - f(x)}{\Delta x} .$$

The important message of the present chapter is that one cannot calculate this, or any other difference quotient, accurately with a calculator when the denominator is "small" except with the aid of ideas from the calculus.

1. Slopes of lines.

A difference quotient is the quotient of two differences. The first difference quotient that occurs in T- F is in Formula (1) in Sect. 1 - 4, namely

$$\frac{\text{rise}}{\text{run}} = \frac{\Delta y}{\Delta x} = \frac{y_2 - y_1}{x_2 - x_1} = m . \tag{1.1}$$

To find the slope, m, of the line through two different points $P_1(x_1, y_1)$ and $P_2(x_2, y_2)$ by this formula, you could use the program

RPN

y_2, [↑], y_1, [-], x_2, [↑], x_1, [-], [÷] .

RPN

AE

y_2, [-], y_1, [=], [÷], [(], x_2, [-], x_1, [=] .

AE

AE
|
AE   or x_2, [-], x_1, [=], [STO], y_2, [-], y_1, [=], [÷], [RCL], [=].

If $x_1 = x_2$, you should not carry out this procedure on the calculator. It objects to being asked to divide by zero. If you try this anyhow, the calculator will indicate an error and stop functioning, and you will have to look in the manual to see how to get it operating properly again. If $x_1 = x_2$, you should properly say that the line has no slope, but it does no harm to indulge in a bit of whimsey and say that the slope is ∞.

If m is the slope of a line, you get the slope of a perpendicular line by taking the negative reciprocal. With m in the display, this can be done by pressing the reciprocal key and the change sign key, in either order.

Calculus books provide many problems involving the calculation of slopes of lines and of perpendiculars, or determining equations of lines through two points or perpendicular to other lines, or related problems involving difference quotients. If such problems are assigned in your calculus course, work them on your calculator, using one of the programs given above, or something analogous. If none are assigned, work a sampling from the following, which are taken from Sect. 1-4 and Sect. 1-5 of T-F.

Problem 1. 1. Plot the given points A and B, and find the slope (if any) of the line determined by them. Find the slope of a line perpendicular to AB, in each case.

1. $A(1,-2)$, $B(2,1)$ 2. $A(-2,-1)$, $B(1,-2)$

3. $A(1,0)$, $B(0,1)$ 4. $A(-1,0)$, $B(1,0)$

5. $A(2,3)$, $B(-1,3)$ 6. $A(1,2)$, $B(1,-3)$

7. $A(0,0)$, $B(-2,-4)$ 8. $A(\frac{1}{2},0)$, $B(0,-\frac{1}{3})$

Problem 1. 2. In the following problems, plot the points A, B, C, and D. Then determine whether or not ABCD is a parallelogram. Say which parallelograms are rectangles.

1. $A(0,1)$, $B(1,2)$, $C(2,1)$, $D(1,0)$

2. $A(-2,2)$, $B(1,3)$, $C(2,0)$, $D(-1,-1)$

3. $A(-1,-2)$, $B(2,-1)$, $C(2,1)$, $D(1,0)$

4. $A(-1,0)$, $B(0,-1)$, $C(2,0)$, $D(0,2)$

Problem 1. 3. In the following problems, use slopes to determine whether the given points are collinear (lie on a common straight line).

1. $A(1,0)$, $B(0,1)$, $C(2,-1)$

2. $A(-2,1)$, $B(0,5)$, $C(-1,2)$

3. A(-2, 1), B(-1, 1), C(1, 5), D(2, 7)

Problem 1. 4. In each of the following problems (1 through 9), plot the given pair of points and find an equation for the line determined by them.

1. (0, 0), (2, 3)
2. (1, 1), (2, 1)
3. (1, 1), (1, 2)
4. (-2, 1), (2, -2)
5. (-2, 0), (-2, -2)
6. (1, 3), (3, 1)
7. (0, 0), (1, 0)
8. (0, 0), (0, 1)
9. (2, -1), (-2, 3)

Problem 1. 5.

a) Find the line L through A(-2, 2) and perpendicular to the line L' : $2x + y = 4$.

b) Find the point B where the lines L and L' of part (a) intersect.

c) Using the result of part (b), find the distance from the point A to the line L' of part (a).

Suppose you have a curve $y = f(x)$. Take different points $P_1(x_1, f(x_1))$ and $P_2(x_2, f(x_2))$ on the curve. Then the slope of the secant joining those two points is the difference quotient

$$\frac{f(x_2) - f(x_1)}{x_2 - x_1} \tag{1.2}$$

of the function f at the values x_1 and x_2. If you hold x_1 fixed and let x_2 approach x_1, this slope approaches the slope of the curve at P_1.

For example, in Sect. 1 - 7 of T - F, it is shown that if

$$f(x) = x^3 - 3x + 3,$$

then the slope at the point $P(x, f(x))$ is $3x^2 - 3$. Thus, the slope at P(2, 5) is 9. Let us take $x_1 = 2$, and $x_2 = 2 + \Delta x$ for various values of Δx, and see if the slope appears to be approaching 9 as $\Delta x \to 0$. This is shown in Table 1. 1. (Note. Some of the final zeros are not significant, but were merely copied from the display in the calculator. These calculations were done on the HP-33E. At the end, a ten digit display was got by executing [MANT], which must be actuated by executing [g].)

At first sight, you might think that Table 1. 1 PROVES that the slope is really 9, since the two values 10^{-5} on either side of $x = 2$ give 9 as the slope. In the first place, no calculation performed at a point different from $x = 2$ can PROVE anything about what happens at $x = 2$, as will be made evident in the next few pages. More to the point, Table 1. 1 is misleading. Though the calculator shows 9 zeros to the right

Table 1. 1

Δx	calculated slope of secant
1	16. 000 00000
0. 1	9. 6100 00000
0. 01	9. 0601 00000
0. 001	9. 0060 01000
0. 0001	9. 0006 00000
0. 00001	9. 0000 00000
-0. 00001	9. 0000 00000
-0. 0001	8. 9994 00000
-0. 001	8. 9940 01000
-0. 01	8. 9401 00000
-0. 1	8. 4100 00000
-1	4. 0000 00000

of the "9" in the entry corresponding to $\Delta x = 0.00001$, leading one to think that the slope of the secant is 9 to 10 significant digits, in fact only the first 4 of these zeros are significant. When Δx gets close to 0, the calculator gives erroneous values, for reasons which will be explained in the next section. In Table 1. 2 we give the exact values of the slope of the secant for those values of Δx for which the values shown in Table 1. 1 are incorrect.

Table 1. 2

Δx	true slope of secant
0. 0001	9. 0006 00010 00000
0. 00001	9. 0000 60000 10000
-0. 00001	8. 9999 40000 10000
-0. 0001	8. 9994 00010 00000

It still appears reasonable that the slope of the curve at $x = 2$ is 9.

Problem 1. 6. In Example 6 in Sect. 1 - 8 of T- F, it is shown that the slope of the curve $y = \sqrt{x}$ at $x = 4$ is 0. 25. Find the slopes of the secants connecting $P_1(4, \sqrt{4})$ and $P_2(4 + \Delta x, \sqrt{4 + \Delta x})$ for $\Delta x = \pm 1, \pm 0.1, \pm 0.01, \pm 0.001, \pm 0.0001$, and ± 0.00001, and see if this seems to substantiate that the slope at $x = 4$ is 0. 25.

Remark. If you are going to store a program for this problem on a programmable calculator, you might as well make it take care of changing the values of Δx. This can be done as follows. Assign a register to store the various values of Δx, and initially put 10 in that register. Now write and store the first half of the program to do the following. Divide what is stored for Δx by 10, and put the quotient back into the Δx register. Then continue with the second half of the program by calculating the slope of the secant, recalling Δx from the Δx register whenever it is needed for the calculation. Then stop. The first time you run the program, you get the slope for $\Delta x = 1$. The next time, you get the slope for $\Delta x = 0.1$. The next time, you get the slope for $\Delta x = 0.01$. And so on. For the negative Δx's, start by storing -10 in the Δx register.

Problem 1.7. It will turn out that the slope of the curve $y = \sin x$ at $x = \pi/3$ is 0.5. Find the slopes of the secants connecting

$$P_1\left(\frac{\pi}{3}, \sin\frac{\pi}{3}\right) \text{ and } P_2\left(\frac{\pi}{3} + \Delta x, \sin\left(\frac{\pi}{3} + \Delta x\right)\right)$$

for $\Delta x = \pm 1, \pm 0.1, \pm 0.01, \pm 0.001, \pm 0.0001$, and ± 0.00001, and see if this seems to substantiate that the slope at $x = \pi/3$ is 0.5. (See what happens if you neglect to use radians. Explain why the numbers you get might have been expected.)

Problem 1.8. For

$$f(x) = \frac{x}{x+1}$$

(see Prob. 8 at the end of Sect. 1-8 of T-F) the formula worked out for $f'(x)$ gives $f'(1) = 0.25$. So, by Formula (3) in Sect. 1-8 of T-F,

$$\frac{1}{4} = f'(1) = \lim_{\Delta x \to 0} \frac{f(1 + \Delta x) - f(1)}{\Delta x} .$$

Calculate the difference quotient

$$\frac{f(1 + \Delta x) - f(1)}{\Delta x}$$

for $\Delta x = \pm 0.1, \pm 0.01, \pm 0.001, \pm 0.0001$, and ± 0.00001, and compare with $f'(1)$.

Problem 1.9. The relationship between demand and price for coffee in the USA from 1960 through 1974 was given by

$$Ad + Bp = 1. \tag{1.3}$$

Here A and B are constants, d is the number of pounds consumed per person per year, and p is the price in dollars per pound. (See Prob. 19 at the end of Sect. 1-5 of T-F.) Suppose that the demand would be 20 pounds per person per year if coffee were given away free, while nobody would drink any coffee at all if the price should rise to \$3.35 per pound. What are the values of A and B?

2. Danger of cancellation.

Suppose we continue Prob. 1. 6 down to $\Delta x = \pm 10^{-7}$, $\pm 10^{-8}$, and beyond. We get the results shown in Table 2. 1. (Again, we copy zeros from the display even though some of them are not significant.)

Table 2. 1

Δx	calculated slope of secant		
	on HP-33E	on TI-57	on TI-SR 50A
10^{-7}	0. 25000	0. 24000	0. 25001
10^{-8}	0. 20000	0. 20000	0. 25010
10^{-9}	0. 00000	0. 00000	0. 25100
10^{-10}	0. 00000	0. 00000	0. 27000
10^{-11}	0. 00000	0. 00000	1. 00000
10^{-12}	0. 00000	0. 00000	1. 00000
10^{-13}	0. 00000	0. 00000	0. 00000
-10^{-13}	0. 00000	0. 00000	0. 00000
-10^{-12}	0. 00000	0. 00000	0. 00000
-10^{-11}	0. 00000	0. 00000	0. 20000
-10^{-10}	0. 00000	10. 00000	0. 24000
-10^{-9}	0. 00000	1. 00000	0. 25000
-10^{-8}	0. 30000	0. 30000	0. 25000
-10^{-7}	0. 25000	0. 26000	0. 25000

We were doing fine until Δx got very close to 0, and then things went all to pieces. On any other calculator, similar results would be forthcoming. Indeed, for an 8-digit calculator, you would probably already be getting peculiar answers at $\Delta x = \pm 10^{-6}$, and if your calculator carries still fewer digits you would be in trouble still sooner.

Or consider the curve $y = x^{-1}$. At $x = 0.7$, the slope of this, to 10 significant digits, is seen to be

$$\frac{-1}{0.49} \cong -2.0408\ 16327 \ .$$

Let us calculate the slopes of the secants connecting the points at $x = 0.7$ and at $x = 0.7 + \Delta x$ for $\Delta x = \pm 10^{-n}$, for various values of n. The results are given in Table 2. 2.

Again, we start off very well, but when Δx gets quite small and we should be getting close to the right answer, everything goes wrong. We get better and better down to $\Delta x = 10^{-5}$ or $\Delta x = 10^{-6}$, after which we get worse and worse. On other calculators,

2. Danger of cancellation

Table 2.2

Δx	calculated slope of secant		
	on HP-33E	on TI-57	on TI-SR 50A
10^{-2}	-2.0120 725	-2.0120 7243	-2.0120 72434 6
10^{-3}	-2.0379 05	-2.0379 050	-2.0379 05034
10^{-4}	-2.0405 3	-2.0405 25	-2.0405 2482
10^{-5}	-2.0408	-2.0407 8	-2.0407 872
10^{-6}	-2.0410	-2.0408	-2.0408 13
10^{-7}	-2.0500	-2.0410	-2.0408 2
10^{-8}	-2.1000	-2.0400	-2.0408
10^{-9}	-2.0000	-2.0000	-2.0410
10^{-10}	-10.000	-2.0000	-2.0400
10^{-11}	0.0000	0.0000	-2.0000
10^{-12}	0.0000	0.0000	-2.0000
10^{-13}	0.0000	0.0000	0.0000
-10^{-13}	0.0000	0.0000	-10.000
-10^{-12}	0.0000	0.0000	-3.0000
-10^{-11}	0.0000	-10.000	-2.1000
-10^{-10}	0.0000	-3.0000	-2.0500
-10^{-9}	-2.0000	-2.1000	-2.0410
-10^{-8}	-2.0000	-2.0500	-2.0409
-10^{-7}	-2.0400	-2.0410	-2.0408 2
-10^{-6}	-2.0400	-2.0409	-2.0408 19

similarly erratic results would be forthcoming.

To see how this happens, write out the calculation for $\Delta x = 10^{-8}$, as done on the HP-33E, which is a 10 digit calculator. We get

$$\frac{1.4285\ 71408 - 1.4285\ 71429}{10^{-8}} = -2.1 .$$

The values of $f(x) = x^{-1}$ at $x = 0.7$ and $x = 0.7 + 10^{-8}$ are very close together, so close that the first 8 digits agree. When we subtract, these first 8 digits cancel, leaving us only a 2 digit answer.

III. DIFFERENCE QUOTIENTS

Let us look at the precisions and accuracies involved in this calculation. The HP-33E gave the approximations

(2.1) $$1.4285\ 71408$$

(2.2) $$1.4285\ 71429$$

for $(0.7 + 10^{-8})^{-1}$ and $(0.7)^{-1}$, respectively. These are both precise to 10 digits, but their difference is precise to only 2 digits.

The absolute error for (2.1) is about 1.6×10^{-10}, and that for (2.2) is about -4.3×10^{-10}. When we subtract, the absolute errors should subtract. So the absolute error for the difference should be about 5.9×10^{-10}, which it is indeed. So our subtraction has not made a great difference in the absolute error.

The relative error for (2.1) is about 1.1×10^{-10}, and that for (2.2) is about -3.0×10^{-10}. However, when we subtract, we get a very much greater relative error for the difference, namely about 2.9×10^{-2}.

To get the slope of the secant, we must now divide the difference of (2.1) and (2.2) by 10^{-8}. This leaves the relative error unchanged, so that it is still about 2.9×10^{-2}. However, it multiplies the absolute error by 10^8, increasing it to about 5.9×10^{-2}.

So, if we take two points with nearly equal x-coordinates, and try routinely on the calculator to calculate the difference quotient or the slope of the secant, we will have bad cancellation, and will come out with an answer that is neither very precise nor very accurate.

We can calculate a much more precise and accurate answer by using a calculus trick. For $f(x) = x^{-1}$,

$$\frac{f(x+\Delta x) - f(x)}{\Delta x} = \frac{\frac{1}{x+\Delta x} - \frac{1}{x}}{\Delta x} = \frac{1}{\Delta x}\,\frac{x-(x+\Delta x)}{(x+\Delta x)\,x} = \frac{-1}{(x+\Delta x)\,x}.$$

That is, the difference quotient can be put in the form

(2.3) $$\frac{-1}{(x+\Delta x)x}$$

where there is no danger of cancellation.

If we put $x = 0.7$ and $\Delta x = 10^{-8}$ in (2.3), we get on the HP-33E the correct slope to 10 digits, namely

$$-2.0408\ 16297\ .$$

But if we just subtract without using the calculus trick, we lose 8 digits, and cannot hope to be anywhere near right.

2. Danger of cancellation

Moral. If one wishes to calculate a difference quotient for two points very close together, one had better use some ideas from calculus to avoid a poor result, i. e., to avoid cancellation.

Problem 2.1. Find the appropriate calculus trick which would allow you to calculate the difference quotient $(f(x + \Delta x) - f(x))/\Delta x$ for f accurately, in case

(a) $f(x) = x^2$

(b) $f(x) = x^4$

(c) you already know how to calculate the difference quotient for the functions g and h accurately and

1. $f(x) = g(x)h(x)$ 2. $f(x) = g(h(x))$.

Hint. Look at how your calculus book derives the formulas for differentiating such a function f (for (c)2, look up the derivation of the "chain rule").

We discuss further ways of using ideas of calculus to combat cancellation when evaluating $f(x + \Delta x) - f(x)$ for "small" Δx in Sect. 3 of Chap. IV.

Chapter IV

SOURCES OF ERROR

0. Guide for the reader.

Almost every calculation performed on a calculator is subject to some error, for reasons which will be explained during the course of the chapter. Manufacturers of calculators try to design them so that the error in a single step "doesn't make any difference." Some manufacturers take more pains with this than others. But even with the best calculators, if one has an extended sequence of steps then the errors of the individual steps (each by itself perhaps too small to matter) can accumulate to produce at times a surprisingly large total error. For calculators whose manufacturers have not kept the errors for individual steps as small as they should, this accumulation of error can lead to very misleading final answers without the user being aware anything is wrong.

The most common source of error is roundoff. This is discussed in Sect. 1. The effect of roundoff error can be magnified by cancellation, as discussed in Sect. 2. Possible ways to reduce cancellation are discussed in Sect. 3. Some of these topics involve ideas from differentiation. So it is best to wait until you have some familiarity with differentiation before reading this chapter. However, you should not wait past that point, since it is important to learn how to minimize errors.

1. Roundoff.

Your first brush with roundoff occurs when you try to enter a number into your calculator which won't fit. For example, there is no way to enter the number π into your calculator. The reason for this is that this number cannot be written exactly as a decimal fraction with eight or ten digits or thirteen or a hundred. Now you may argue that your calculator actually boasts a key marked π which supposedly puts the number π into the display every time you press that key. But, it clearly cannot get π into the display to more digits than the display can hold, nor can the calculator store π internally to more digits than the calculator will hold. Some users discover this discrepancy when they use their calculator to compute $\sin \pi$ (in radian mode) and get a nonzero answer. For example, $\sin \pi = -4.1 \times 10^{-10}$ on a certain 10 digit calculator, corresponding to the fact that, on that calculator, the π key delivers the number

1. Roundoff

(1.1) $$3.1415\ 92654$$

rather than the nonterminating decimal expansion

$$\pi = 3.1415\ 92653\ 58979\ 32384\ldots\ .$$

On that calculator, the manufacturer has provided the value of π correctly rounded to 10 decimal digits. This means that the value (1.1) of π provided is the closest number to π obtainable if one has only ten decimal digits to work with.

More generally, <u>to round a number</u> x <u>to</u> N <u>digits</u> means to determine the N-digit number which is closest to x. You can usually obtain this rounded number by writing down the first $N+1$ significant digits of x, adding 5 in the $(N+1)$st place, and then throwing away the $(N+1)$st digit. For example, rounding π to 4 digits gives

$$\begin{array}{r} 3.1415 \\ +\quad\quad 5 \\ \hline 3.1420 \end{array}$$

which gives 3.142, while rounding it to 13 digits gives

$$\begin{array}{r} 3.1415\ 92653\ 5897 \\ +\quad\quad\quad\quad\quad\quad 5 \\ \hline 3.1415\ 92653\ 5902 \end{array}$$

which gives 3.1415 92653 590. So, rounding does not affect only the last digit retained, but may also affect earlier digits. As an extreme example, rounding the number 1.9999 99999 93 to ten digits gives the number 2.0000 00000, all of whose digits are different from the corresponding digits of the original number. One minor exception would occur if one wishes to round 0.99999 99999 7 to ten digits. After adding 5×10^{-11}, you throw away the last <u>two</u> digits to get the rounded ten digit number 1.0000 00000. But rounding 0.99999 99999 3 would give 0.99999 99999.

There is a lazy way to fit a number requiring more than N digits into an N-digit calculator, called <u>truncating</u>. In this way, you simply throw away all the digits after the N-th. But this is worse than rounding for two reasons. It doesn't always get the best N-digit approximation into the calculator, so the error made may be larger than necessary. Also, errors are less apt to offset each other in subsequent calculations, since all numbers start out being less than or equal to what they ought to be (in absolute value). In short, truncating is a <u>biased</u> way to fit numbers into the calculator.

Whether it is done by rounding or truncating or by yet another way, the error made when trying to fit a number into an N-digit calculator is called <u>roundoff error</u>.

<u>Problem 1.1.</u> Round each of the following numbers to 10 digits.

$$\frac{2}{3} = 0.66666\ 66666\ 66666\ldots$$

$$\ln 2 = 0.69314\ 71805\ 59945\ldots$$

$$\pi = 3.1415\ 92653\ 58979\ldots$$

$$\sqrt{2} = 1.4142\ 13562\ 37309\ldots$$

$$\sin 1 = 0.84147\ 09848\ 07896\ldots$$

$$2^{49} = 56294\ 99534\ 21312$$

$$\tan(1.5612\ 91773) = 105.20\ 95563\ 04960\ldots$$

If a number is rounded to N digits, and the rounded value is in turn rounded to M digits, with $M < N$, one will not necessarily get the same value as if the original number had been rounded straight off to M digits.

Problem 1.2. Invent a 6-digit number B such that, if one first rounds it to 4 digits and then rounds that result to 2 digits, one will get a different number from what one gets if one rounds B directly to 2 digits.

Once you begin to operate on the numbers you have somehow managed to get into your N-digit calculator, you incur further roundoff error because the results that you should get by your calculations can usually not be written exactly as N-digit numbers. For example, try dividing 2 by 3. But, calculators differ substantially in how they cope with this problem. We now illustrate this with two calculators, the HP-33E and the TI-57.

The HP-33E is a 10-digit calculator. But, in order to find out exactly what answers it calculates, one has to learn some of its eccentricities. Although it is a 10-digit calculator, the display often shows fewer than 10 digits. It can be set to show very few digits in the display, if one wishes (see the manual), but it will still be holding 10 digits internally. At any time, one can see what the 10 internal digits are by pressing the MANT key (which has to be actuated by first pressing the key g). If one presses SCI followed by 7, 8, or 9, the display will show a 7-digit truncated approximation to what the calculator is holding internally. If one presses SCI followed by 6, the display will show a 7-digit rounded approximation to what the calculator is holding internally.

The TI-57 is an 11 digit calculator, but the display never shows more than 8 digits, which are a rounded version of what the calculator is holding internally. As explained in the manual, one can set the display to show fewer digits. However, often the 8-digit display is actually giving fewer significant digits, because there are some zeros in the lefthand spaces.

For example, if one divides 2 by 3, the display shows 0.6666 667. The calculator is actually holding

$$0.66666\ 66666\ 6$$

internally, but to find this out one has to use some trickery. After dividing 2 by 3, subtract 0.66666. Then the display shows 0.0000 067. However, the calculator is then actually holding

0.00000 66666 6

internally. To show this, multiply by 100,000, whereupon the display shows 0.6666 66.

Incidentally, one cannot key a number of more than 8 digits directly into the display of the TI-57. If one attempts to key in 1.5612 91773 directly by pressing the digits (and decimal point) in order, the TI-57 will accept the first 8 digits (with the decimal point) and show them in the display, but will completely ignore the final 7 and 3. Although the TI-57 can actually hold 11 digits, it refuses to accept more than 8 if they are just keyed in successively. In order to get 1.5612 91773 into the TI-57, you first key in 9.1773×10^{-5}, and then add 1.5612 to it.

Still other calculators have still other variations of this eccentricity.

Problem 1.3. Find out how many digits your calculator carries internally, and how to read all of them out. Also, find out how to enter as many digits of a number as possible into your calculator.

Problem 1.4. Each of the numbers mentioned in Prob. 1.1 is obtainable (approximately) on your calculator with a few key strokes. Use what you have learned in Prob. 1.3 to find out how accurately your calculator provides the numbers mentioned in Prob. 1.1. If you have an 8-digit or 9-digit calculator, replace the last line of Prob. 1.1 by

tan (1.5612 918) = 105.20 98551 97179 . . . .

Incidentally, if you should try Prob. 1.4 on an HP-33E, you would get the 10-digit rounded numbers you were supposed to get for Prob. 1.1. That is, the approximations given by the HP-33E are as close to the numbers as anyone can get with 10 digits.

You do not do this well with every calculator, as you possibly found out when you tried Prob. 1.4. For example, the answers produced by the TI-57 are given in Table 1.1, together with their errors.

Of these approximations, only the one for π is correctly rounded to 11 digits. The next to last number is so far off that the error is already evident from the rounded 8 digit display, which is all that calculator will let you see without a special effort on your part.

We consider now in more detail how these two calculators cope with the fact that the result of a calculation does not fit into the calculator. Consider, for example, the addition of two numbers of "different" sizes. Say we wish to add

$$\begin{array}{r} 10000\ 00000 \\ +\ 666\ 66666.67\ . \end{array}$$

Table 1. 1

expression	as evaluated on TI-57	approximate error
2/3	0. 66666 66666 6	6.7×10^{-12}
ln 2	0. 69314 71810 0	$-440. \times 10^{-12}$
π	3. 1415 92653 6	-1.0×10^{-11}
$\sqrt{2}$	1. 4142 13562 0	$37. \times 10^{-11}$
sin 1	0. 84147 09853 8	$-570. \times 10^{-12}$
2^{49}	5. 6294 99652 8×10^{14}	$-12000. \times 10^{3}$
tan (1. 5612 91773)	105. 20 95585 2	$-2200. \times 10^{-9}$

This would give a 12-digit sum, which is beyond the capacity of either the HP-33E or the TI-57. In such a case, the TI-57 truncates all digits in the smaller number beyond what would make 11 digits for the larger number. So it gives the sum

10666 66666. 6 .

For the subtraction

```
  1 00000 00000
-   66666 66666. 7 ,
```

it truncates the second number, giving

```
  1 00000 00000
-   66666 66666  ,
```

and so gets

33333 33334

for the difference.

The HP-33E apparently adds an extra zero to the larger number, making it temporarily an 11-digit number. It then rounds the smaller number so that it does not stick out beyond that. Then it adds (or subtracts) and, if 11 digits result, it rounds down to 10. Thus, for the first addition, the HP-33E gives 10666 66667. For the subtraction

```
  1. 0000 00000
- . 6666 66666 7 ,
```

it gives

0. 33333 33333 .

1. Roundoff

Next, we consider multiplication. When a calculator that holds N digits is asked to calculate the product of two N-digit numbers which it is holding, it generates a product with 2N digits (sometimes only 2N - 1). This must be reduced to N digits so the calculator can hold an approximation for the product. Some calculators do this by rounding off the full product, and some by truncating it. Rounding off is preferable, of course.

The HP-33E rounds off after multiplying. Thus, the product of 1.1111 112 and 1.1111 111 is

$$1.2345\ 67987\ 65432\ .$$

For this, the HP-33E gives the approximation

$$1.2345\ 67988,$$

correctly rounded to 10 places.

The TI-57 truncates after multiplying. It gives the product above internally as

$$1.2345\ 67987\ 6,$$

truncated to 11 places. It rounds this correctly to 8 digits for the display.

Problem 1.5. Try to find out if your calculator rounds or truncates after multiplying.

We hope that you are by now convinced that your calculator usually makes an error when carrying out any of the arithmetic operations and that these errors differ from calculator to calculator, even if you start with the same numbers.

There are also errors associated with the use of the various function keys on your calculator, such as the $\sqrt{x}$ key or the sin key and the like. We comment on this in Chap. I . To be precise, such a key delivers a certain function exactly and without error. But, you are usually not interested in the function given by the key, but rather in the function written on the key, and these two functions usually differ. If, for example, a certain 10-digit calculator delivers the answer

$$1.4142\ 13562$$

when you press the $\sqrt{x}$ key with the number 2 in the display, then this answer is, as an approximation to the number $\sqrt{2}$, in error by about 0.373×10^{-9}

The error in an answer produced by one of the calculator keys depends on the argument and also varies from calculator to calculator even if you start with the same argument. For example, the sin key delivers the answer 0.84147 09848 on an HP-33E and the answer 0.84147 09853 8 on a TI-57 when the number 1 is in the display (and the calculator is in radian mode). For comparison, sin 1 = 0.84147 09848 07896... . This also shows that the error is often more than just the roundoff due to the fact that the correct answer won't fit into the calculator.

In short, if a key on your calculator purports to evaluate the calculus function f, and if you enter a number x into the display and press the key, you will get

$$f(x) - \epsilon$$

instead of the exact value f(x). Here ϵ is the inherent calculator error. It varies from one function to another, it varies from one x to another, and it varies from one calculator to another. We just saw an instance for the function sin. Let us see what happens for tan. In Table 1. 2 we list five values of x, together with values of their tangents

Table 1. 2

	x	tan x
x_1	9. 5045 53935 $\times 10^{-3}$	9. 5048 40148 20 $\times 10^{-3}$
x_2	1. 5612 91772	105. 20 95452 35
x_3	1. 5612 91772 4	105. 20 95496 63
x_4	1. 5612 91773	105. 20 95563 05
x_5	102. 09 19665	102. 09 23322 61

rounded to 12 significant digits. Values of the tangents, with errors, as calculated on two calculators are given in Table 1. 3 for the same values of x that were used in

Table 1. 3

x	tan x			
	on HP-33E	ϵ	on TI-57	ϵ
x_1	9. 5048 40148 $\times 10^{-3}$	2×10^{-13}	9. 5048 40007 1 $\times 10^{-3}$	1411×10^{-13}
x_2	105. 20 95452	35×10^{-9}	105. 20 95474 6	-2225×10^{-9}
x_3			105. 20 95518 8	-2217×10^{-9}
x_4	105. 20 95563	5×10^{-9}	105. 20 95585 2	-2215×10^{-9}
x_5	102. 09 23322	61×10^{-9}	102. 09 23643 6	-32099×10^{-9}

Table 1. 2, except that we could not give a value on the HP-33E for x_3 since x_3 is an 11-digit number and would not fit into the display of the HP-33E.

The same discussion applies to a function that cannot be approximated by pressing a single key, but requires a succession of key strokes. That is, you would have to write a program to evaluate f. Every key that you press in the program contributes an error, and these errors accumulate in a complicated fashion. So, if you enter a number x into the display and run your program, you will get

$$f(x) - \epsilon$$

instead of the true value $f(x)$.

If you use the same program to calculate $f(x')$ for some $x' \neq x$, you obtain the number

$$f(x') - \epsilon'$$

and the error ϵ' is, in general, different from ϵ. But, typically, for x' "near" x, you should expect all errors ϵ' to fall into some interval $(\epsilon_{min}, \epsilon_{max})$. Under ideal circumstances, the calculated value for $f(x)$ is obtained by rounding the exact value $f(x)$ to the number of digits carried by the calculator. In that case,

$$\epsilon_{max} = -\epsilon_{min} = 5 \times 10^{-d-1} |f(x)| ,$$

assuming the calculator to carry d digits. Usually, ϵ_{max} and $-\epsilon_{min}$ will be larger than that, but usually $\epsilon_{max} = -\epsilon_{min}$. It is possible, though, to have ϵ_{max} quite different from $-\epsilon_{min}$, in the presence of <u>systematic</u> error in the calculated function values. In any case, we refer to the larger of the two numbers ϵ_{max} and $-\epsilon_{min}$ as the <u>noise level</u> in the calculated function values. Thus,

$$|\text{calculated value of } f \text{ at } x - f(x)| \leq \text{noise level} .$$

For example, we would surmise from Table 1.3 that the noise level in function values $\tan x$ for x near 1.5612 91772 is about 4×10^{-8} on an HP-33E and about 2×10^{-6} on a TI-57. Of course, more function values would be required to come to a reliable statement about the noise level. Also, one would note that, on the TI-57, all errors for x near 1.5612 91772 agree to about three digits. In fact

$$(\epsilon_{min}, \epsilon_{max}) \cong (-223 \times 10^{-8}, -221 \times 10^{-8}),$$

which is a sign of <u>systematic error</u> in the algorithm used for tan on that calculator.

It is usually quite tricky if not impossible to understand how the errors made at each step in your program for f accumulate to give the error ϵ in the final answer for $f(x)$. As an indication of what might be involved in the study of such <u>error propagation</u>, we now discuss the situation where we are supposed to evaluate the function f at some point x_0, but have available only the approximation $x_0 - \Delta x$ to x_0. The question is: How does this argument error Δx affect the accuracy of the computed answer for $f(x_0)$?

The computed answer

$$f(x_0 - \Delta x) - \epsilon'$$

is now in error on two counts. There is the calculation error ϵ' and there is also the error $f(x_0) - f(x_0 - \Delta x)$. This latter error we can gauge with the help of calculus.

Looking forward to (1. 4) of Chap. VI, we have

$$f(x_0 - \Delta x) \cong f(x_0) - f'(x_0)\Delta x .$$

So the total error in our computed answer is about

$$f'(x_0)\Delta x + \epsilon' .$$

(Recall that error = true value - approximation.) In other words, the error Δx in the argument $x_0 + \Delta x$ enters the final answer multiplied by $f'(x_0)$. Hence, the greater $|f'(x_0)|$, the more <u>sensitive</u> is the final answer to the error Δx in the argument $x_0 + \Delta x$.

To be precise, Δx has also affected the calculation error, since, as we said earlier, we cannot expect ϵ' to equal the error ϵ associated with the argument x_0. But, though ϵ and ϵ' could be considerably different, each is less than the noise level in absolute value. In summary, we expect the calculated value for $f(x_0)$ to be of the form

$$f(x_0) - f'(x_0)\Delta x - \epsilon' , \tag{1. 2}$$

with $|\epsilon'| \leq$ noise level.

We illustrate these ideas, using once again the tangent. Suppose you wish to calculate $\tan(1.104\sqrt{2})$. From the accurate value of $\sqrt{2}$ given in Prob. 1. 1, you can calculate that to 14 significant digits

$$1.104\sqrt{2} \cong 1.5612\ 91772\ 8599 .$$

This has too many digits to input in either the HP-33E or the TI-57, or most any other calculator. Of course, you can round it off to the nearest number that can be input. For the HP-33E, this would be

$$1.5612\ 91773 . \tag{1. 3}$$

For the TI-57, this would be

$$1.5612\ 91772\ 9 . \tag{1. 4}$$

Actually, the situation is worse than that. You start off innocently to calculate $\tan(1.104\sqrt{2})$. So you take $\sqrt{2}$ and multiply by 1. 104, and think you are ready to press the tan key. But what do you have in the display? On the HP-33E, you have not (1. 3) but

$$1.5612\ 91772 . \tag{1. 5}$$

On the TI-57, you have not (1. 4) but

$$1.5612\ 91772\ 4 . \tag{1. 6}$$

In order to determine the error in the numbers (1. 5) and (1. 6) as approximations to $1.104\sqrt{2}$, we use a more precise approximation for $\sqrt{2}$ than given in Prob. 1. 1, and find that

(1. 7) $$1.104\sqrt{2} = 1.5612\ 91772\ 85989\ldots\ .$$

We have

(1. 8) $$\tan(1.104\sqrt{2}) = 105.20\ 95547\ 54012\ldots\ .$$

We have

$$\frac{d}{dx}\tan x = \sec^2 x\ .$$

So for $f(x) = \tan x$, we have for $x_0 = 1.104\sqrt{2}$

$$f'(x_0) \cong 11070\ .$$

On the HP-33E, we got

$$x_0 - \Delta x = 1.5612\ 91772$$

(see (1. 5)). This is the x_2 of Table 1. 2. So

$$\Delta x \cong 8.5989 \times 10^{-10}\ .$$

Hence

$$f'(x_0)\Delta x \cong 95 \times 10^{-7}\ ,$$

which is much larger than the noise level 4×10^{-8} surmised from Table 1. 3. So, in this case, the error in the HP-33E value for $\tan x_2$ (recorded in Table 1. 3), if considered as an approximation to $\tan(1.104\sqrt{2})$ (given in (1. 8)), is almost entirely due to the discrepancy Δx between x_2 and $x_0 = 1.104\sqrt{2}$.

On the TI-57, we got

$$x_0 - \Delta x = 1.5612\ 91772\ 4$$

(see (1. 6)). This is the x_3 of Table 1. 2. So

$$\Delta x \cong 4.5989 \times 10^{-10}\ .$$

Hence

$$f'(x_0)\Delta x \cong 5091 \times 10^{-9}\ .$$

This is a bit over twice the noise level which we surmised earlier from Table 1. 3. In this case, the noise level is therefore a significant part of the error. More explicitly, we find, with ϵ as taken from Table 1. 3 for x_3, that

$$f'(x_0)\Delta x + \epsilon \cong (5091 - 2217) \times 10^{-9} = 2874 \times 10^{-9} .$$

This agrees to all four digits with the error got by subtracting the value of $\tan x_3$ in Table 1. 3 from the value given above for $\tan(1.104\sqrt{2})$.

<u>Problem 1. 6.</u> Show that you may expect difficulties in evaluating expressions of the sort

$$(f(x))^n$$

accurately for large n.

The only simple remedy against roundoff error is to carry more digits in the calculations. Since the number N of digits which your calculator carries is fixed, this requires special procedures. A common scheme is <u>double precision</u> in which all calculations are carried to $2N$ digits. For each number, the $2N$ most significant digits are stored in <u>two</u> registers. Procedures are worked out to add, subtract, multiply and divide two such numbers with the aid of the ordinary arithmetic of the calculator. On the more sophisticated programmable calculators, there are programs available for this. But you would find it quite a challenge to develop such programs yourself.

<u>Problem 1. 7.</u> For the argument 1. 5612 91773 of the tan in Prob. 1. 1, use the accurate value of π provided in Prob. 1. 1 to compute the first ten significant digits of the difference d between $(\pi/2)$ and said argument. Then use this accurate value to calculate

$$\tan(1.5612\ 91773) = \tan(\frac{\pi}{2} - d) = \cot d = \frac{1}{\tan d}$$

on your calculator and compare with the value given in Prob. 1. 1 and in Table 1. 1.

<u>Problem 1. 8.</u> Use an accurate value of $\pi/2$ with the value of $1.104\sqrt{2}$ given in (1.7) to calculate an accurate value of

$$D = \pi/2 - 1.104\sqrt{2} .$$

Then

$$\tan(1.104\sqrt{2}) = \tan(\pi/2 - D) = \cot(D) = 1/\tan(D) .$$

Use this to calculate an accurate value of $\tan(1.104\sqrt{2})$, and compare with (1. 8).

2. Cancellation.

Cancellation occurs when you subtract a number from another which is almost equal to it. For example, let us calculate the difference $22/7 - \pi$ on a 10-digit calculator. This gives

$$\begin{array}{r} 3.1428\ \ 57143 \\ -\ \underline{3.1415\ \ 92654} \\ 0.0012\ \ 64489 \end{array}$$

Note that we incurred no roundoff error in this subtraction. So, what is the problem?

The problem becomes apparent when you compare our computed answer with the accurate difference

$$22/7 - \pi = 0.0012\ 64489\ \ 26734\ldots\ .$$

You now see that our computed difference, as an approximation to the number $22/7 - \pi$, is accurate only to seven digits, even though we started with approximations to $22/7$ and π which are accurate to 10 digits. We lost three digits of accuracy, because we lost three digits of precision. The first three significant digits in the two numbers $22/7$ and π coincide and therefore cancel each other when we subtract.

In other words, cancellation is not an error in itself. Rather, cancellation allows earlier errors to become more prominent.

Cancellation can happen in less obvious ways. For instance, consider the polynomial

$$x^4 - 8x^3 + 12x^2 + 16x + 4\ . \tag{2.1}$$

Let us try to calculate its value for $x = 40/9$. Arranged for Horner's method (see Sect. 1 of Chap. II), the polynomial has the form

$$(((x - 8)x + 12)x + 16)x + 4\ .$$

With $x = 40/9$, we have on the HP-33E calculator

$$(x - 8)x \cong -15.802\ \ 46914\ ,$$

So when we add 12, the first digit cancels. Then

$$((x - 8)x + 12)x \cong -16.899\ \ 86284\ .$$

So when we add 16, two more digits cancel. Then

$$(((x - 8)x + 12)x + 16)x \cong -3.9993\ \ 90400\ .$$

So when we add 4, four more digits cancel, and we get the calculated value

$$0.00060\ 96000\ .$$

Altogether, we have lost 7 significant digits through cancellation. As we were using the HP-33E, which has 10-digit arithmetic, we cannot expect more than 3 digits to be correct at this stage. The value of the polynomial is

$$\frac{4}{9^4} = 0.00060\ 96631\ 61103\ 49032\ldots\ .$$

So we were a bit lucky, as the fourth significant digit is nearly correct.

Cases of severe cancellation occurred in Sect. 2 of Chap. III. In particular, look a few lines below Table 2. 2 of Chap. III, where the first eight digits cancelled, leaving only <u>two</u> correct.

<u>Problem 2. 1.</u> See what value your own calculator gives for the polynomial (2. 1) when $x = 40/9$.

<u>Problem 2. 2.</u> Attempt to calculate the difference quotient

$$\frac{\sqrt{x + \Delta x} - \sqrt{x}}{\Delta x} \tag{2. 2}$$

for $x = 3$ and $\Delta x = 10^{-7}$.

<u>Remark</u>. Calculated values of (2. 2) for $x = 4$ and various small values of Δx are given in Table 2. 1 of Chap. III.

<u>Problem 2. 3.</u> Attempt to calculate the difference quotient

$$\frac{\sqrt[3]{x + \Delta x} - \sqrt[3]{x}}{\Delta x} \tag{2. 3}$$

for $x = 4$ and $\Delta x = 10^{-7}$.

<u>Problem 2. 4.</u> Attempt to approximate

$$f(x) = x - \ln x - 2 \tag{2. 4}$$

for $x = 3.1462$. How many digits do you think are correct in your calculated answer?

<u>Problem 2. 5.</u> Try to calculate the difference quotient

$$\frac{\sin(x + \Delta x) - \sin(x - \Delta x)}{2\Delta x} \tag{2. 5}$$

for $x = 2$ and $\Delta x = 10^{-7}$.

<u>Problem 2. 6.</u> Calculate

2. Cancellation

(2. 6)
$$\sinh\left(\frac{1}{2^{20}}\right) .$$

You will learn in due course (if you do not already know it) that sinh x is called the hyperbolic sine of x, and is defined as

(2. 7)
$$\sinh x = \frac{1}{2}(e^{x} - e^{-x}) .$$

If your calculator has a sinh key, use (2. 7) nevertheless for the calculation, so as to cause cancellation.

Remark. An accurate approximation for (2. 6) is

$$9.5367\ 43164\ 0625 \times 10^{-7} .$$

A famous equation is

(2. 8)
$$e^{x} = 1 + \frac{x}{1!} + \frac{x^2}{2!} + \frac{x^3}{3!} + \frac{x^4}{4!} + \dots ,$$

which you will learn when you come to series in the calculus. Of course, this is an infinite sum, and so you cannot evaluate it by adding it up. But you will then also learn that

(2. 9)
$$e^{x} = p_N(x) + R_N(x)$$

with p_N the polynomial given by

(2. 10)
$$p_N(x) = 1 + \frac{x}{1!} + \frac{x^2}{2!} + \dots + \frac{x^N}{N!}$$

and

(2. 11)
$$R_N(x) = \frac{x^{N+1}}{(N+1)!} e^{\xi} ,$$

for some ξ between 0 and x. It is possible to verify that

(2. 12)
$$\left| \frac{R_{50}(-10)}{e^{10}} \right| < 5 \times 10^{-11}$$

You would therefore expect to calculate e^{-10} to 10 digit accuracy by evaluating the polynomial p_{50} at -10. And, indeed, by (2. 9) and (2. 12) this would succeed if you could evaluate p_{50} to 10 digit accuracy. But unfortunately, you encounter rather severe cancellation if you attempt to calculate $p_{50}(-10)$ on a calculator.

You could, of course, use Horner's method to evaluate p_{50}. That would give

$$(\ldots((\frac{x}{50!} + \frac{1}{49!})x + \frac{1}{48!})x + \ldots + \frac{1}{1!})x + 1 .$$

But you would be better off to evaluate p_{50} in an alternate "nested form"

$$((\ldots((\frac{x}{50} + 1)\frac{x}{49} + 1)\frac{x}{48} + \ldots)\frac{x}{2} + 1)\frac{x}{1} + 1 .$$

Use of a program with a suitable subroutine will let you carry out this calculation with little effort.

*Problem 2. 7. Calculate an approximation for e^{-10} by evaluating p_{50} at -10. Compare with what your calculator gives for e^{-10}, and see how bad the cancellation effect is.

*Problem 2. 8. In doing Prob. 1. 8, you might decide that instead of calculating D by hand, you could shuffle off the work onto the calculator. So calculate D by calculating $\pi/2$ and $1.104\sqrt{2}$ on your calculator, and subtracting them. Explain why you get such a poor answer for $\tan(1.104\sqrt{2})$ this way.

In extreme cases, cancellation can be so bad that not more than one correct digit could survive. However, roundoff will usually throw the final digit several units off (or even the last two digits). In such a case, the answer being given by the calculator will have nothing whatsoever to do with the correct answer. When this happens, it is another case of the calculator producing "noise." A particularly severe case occurs near the end of Sect. 3 of Chap. VII.

3. Some ways to reduce cancellation.

Roundoff errors are unavoidable and usually matter little, except when they are catapulted into prominence through cancellation. It is therefore important to be on the lookout for cancellation and to try to avoid cancellation if possible.

At times, cancellation can be reduced or avoided merely by rewriting a formula appropriately. A famous instance of this is the quadratic formula

$$x = \frac{-b \pm \sqrt{b^2 - 4ac}}{2a} \tag{3.1}$$

for the two roots of the quadratic equation

$$ax^2 + bx + c = 0 . \tag{3.2}$$

Suppose we wish to solve

$$x^2 - 200x + 1 = 0 . \tag{3.3}$$

By the quadratic formula, (3. 1), the two roots are

$$\frac{200 \pm \sqrt{40000 - 4}}{2} ,$$

or

(3. 4) $$100 \pm \sqrt{9999} .$$

On the HP-33E, we get

$$\sqrt{9999} \cong 99.994\ 99987 .$$

When we use the + sign in (3. 4), we get

(3. 5) $$199.99\ 49999$$

as one root. But when we use the - sign in (3. 4), the first five digits cancel and we get

(3. 6) $$0.005\ 00013 ;$$

because of the way the HP-33E subtracts, we happen to get six significant digits. So we have at most six significant digits of accuracy (or perhaps only five on some calculators).

Of course, six significant digits is more than one usually needs. So why worry? There may be further cancellations in subsequent calculations, and for that reason we would like to maintain high accuracy as much as possible, particularly when this can be done easily, as in the present case. Take the quadratic formula, (3. 1), and multiply top and bottom by

$$- b \mp \sqrt{b^2 - 4ac} .$$

This gives

(3. 7) $$x = \frac{2c}{- b \mp \sqrt{b^2 - 4ac}} .$$

For the equation (3. 3), this would give

$$\frac{2}{200 \mp \sqrt{40000 - 4}}$$

or

(3. 8) $$\frac{1}{100 \mp \sqrt{9999}} .$$

If we use the - sign on the calculator, we get something like (3. 5), but with a LARGE cancellation. But if we use the + sign, we get

$$\frac{1}{199.99\ 49999},$$

which works out to

$$0.00\ \ 50001\ \ 25006.$$

This is the other root, correctly rounded to 10 significant decimal digits.

We have given the procedure we did for the quadratic formula because it can be adapted to other situations. Actually, for the specific equation (3. 3), the matter could be handled more simply as follows. If r_1 and r_2 are roots of (3. 3), then by Cor. 2 for Thm. 2. 1 of Chap. II,

$$(x - r_1)(x - r_2)$$

must divide the left side of (3. 3). As the coefficient of x^2 must be unity, we get

$$x^2 - 200x + 1 = (x - r_1)(x - r_2).$$

Multiplying out the right side of this, we see that

$$r_1 r_2 = 1 .$$

So, as soon as we find that (3. 5) is an approximation for r_1, we can immediately write down

$$r_2 = \frac{1}{r_1} \cong \frac{1}{199.99\ 49999}.$$

This is the same result that we got above, by a different argument.

Problem 3. 1. For the formula (2. 2), find an equivalent formula that does not have cancellation.

Hint. Use the same trick that was used for the quadratic formula.

Problem 3. 2. For the formula (2. 3), find an equivalent formula that does not have cancellation.

Hint. Multiply top and bottom by

$$(x + \Delta x)^{\frac{2}{3}} + (x + \Delta x)^{\frac{1}{3}} x^{\frac{1}{3}} + x^{\frac{2}{3}} .$$

Problem 3. 3. Calculate both roots of

$$x^2 - 683x - 1 = 0$$

to full accuracy.

Problem 3. 4. For the difference quotient (2. 5), find an equivalent formula that does not have cancellation.

Hint. Remember that addition formula for the sine function,

$$\sin(\alpha \pm \beta) = \sin \alpha \cos \beta \pm \sin \beta \cos \alpha ,$$

and make use of the fact that your calculator will evaluate

$$\frac{\sin \Delta x}{\Delta x}$$

accurately for Δx near (but not at) zero. See Chapter VIII for a discussion of this latter point.

Problem 3. 5. For the difference quotient

$$\frac{\cos(\alpha + \Delta x) - \cos(\alpha - \Delta x)}{2\Delta x},$$

find an equivalent formula that does not have cancellation.

Problem 3. 6. Find a different way to write the expression

$$1 - \cos x$$

for x near zero which avoids cancellation.

Hint. Multiply and divide by $1 + \cos x$.

Remark. In view of Problems 3. 1, 3. 2, 3. 4 - 3. 6, it would be instructive to review Prob. 2. 1 in Chap. III.

*Problem 3. 7. Find a way to calculate e^{-10} without the cancellation that you got in Prob. 2. 7, and perform the calculation.

Hint. Recall that

$$e^{-x} = \frac{1}{e^{x}} .$$

So

$$e^{-10} = \frac{1}{e^{10}} .$$

If we take $x = 10$ in (2. 8), we get e^{10} with no danger of cancellation. Furthermore, on a 10-digit calculator, you can stop with the term

$$\frac{10^{43}}{43!},$$

and get e^{10} with 10 significant digits, since by (2. 11)

$$\left|\frac{R_{43}(10)}{e^{10}}\right| < 5 \times 10^{-11} .$$

A second way to avoid cancellation is through a bit of calculus. Cancellation often occurs when a function f is evaluated near a zero, i. e. , near a point x_0 for which $f(x_0) = 0$. In such a case, looking forward to (1. 4) of Chap. VI, we have

(3. 9) $$f(x) \cong f(x_0) + f'(x_0)(x - x_0) = f'(x_0)(x - x_0)$$

with the error in this approximation the smaller the closer x is to x_0.

If now x is given to us near x_0, then there can be cancellation just in calculating $x - x_0$, and the approximation (3. 9) may then be of little use to us. But, sometimes in such a situation, x is (or can be) given explicitly in the form

$$x = x_0 + h$$

and then (3. 9) can be used effectively in the form

(3. 10) $$f(x_0 + h) \cong f'(x_0)h .$$

For example, in Prob. 2. 6, you experienced much cancellation when evaluating $\sinh(2^{-20})$. In this case, you have $f(x) = \sinh x = (e^x - e^{-x})/2$ and $x_0 = 0$, so that $x = h$. So, as soon as you have learned that

$$\frac{d}{dx}e^x = e^x$$

you can use (3. 10) to do much better.

Problem 3. 8. Calculate $\sinh(2^{-20})$ more accurately by means of (3. 10) and compare with the value given after Prob. 2. 6.

Remark. In the case of the hyperbolic sine, it is very easy to do much better than (3. 10). When you come to series, you will learn that

(3. 11) $$\sinh x = (e^x - e^{-x})/2$$

$$= x + \frac{x^3}{3!} + \dots + \frac{x^{2n-1}}{(2n-1)!} + R_n(x) ,$$

with

$$|R_n(x)| \le \frac{(\cosh x)\,|x|^{2n+1}}{(2n+1)!} \quad .$$

This can be used to show that the approximation

$$\sinh x \cong x + \frac{x^3}{3!} = x\left(1 + \frac{x^2}{6}\right) \tag{3.12}$$

has relative error less than 5×10^{-11} for all x with $|x| \le 8.8 \times 10^{-3}$. This approximation should therefore be used in preference to the formula $(e^x - e^{-x})/2$ whenever $\sinh x$ is to be evaluated for this range of x.

It is a bit trickier to avoid cancellation near a zero of f if you <u>do not know</u> the zero. For example, you found cancellation when you tried, in Prob. 2.4, to evaluate the function f given by

$$f(x) = x - \ln x - 2 \tag{3.13}$$

at $x = 3.1462$. In such a case, you will simply have to carry enough digits during the calculation so that in the end you still retain, in spite of the cancellation, the precision you desire. You found in Prob. 2.4 that $\ln(3.1462) \cong 1.1461\ 95375$, while $x - 2 = 1.1462$, hence there is cancellation of five digits when evaluating (3.13) at $x = 3.1462$. Thus, to get $f(3.1462)$ nevertheless to 10 digits, you would have to calculate $\ln(3.1462)$ somehow to 15 digits.

There is some consolation, though. Once you have obtained the number $f(x_0)$ accurately for some x_0 (whether or not x_0 is a zero for f), you can then safely use the approximation

$$f(x_0 + h) \cong f(x_0) + f'(x_0)h \tag{3.14}$$

for "small" h in order to evaluate f with the same accuracy for all $x = x_0 + h$ "near" x_0.

In practice, for example for the function f given by (3.13), one would actually use more sophisticated approximations than (3.14) in order to allow for relatively large values of h. We discussed such an approximation earlier for $\sinh x$. But the basic idea is the same, namely, to carry out the required calculation for just one value of x, carrying as many digits as is necessary to produce the final value to the desired accuracy, in spite of the cancellations along the way. After that, you have an accurate formula for $f(x + h)$ in terms of h, whose evaluation as a function of h involves no undue cancellation.

As a final example, consider the cubic polynomial p given by

$$p(x) = a_0x^3 + a_1x^2 + a_2x + a_3 \tag{3.15a}$$

with

(3.15b) $\quad a_0 = 1, \quad a_1 = -2.1213\ 20344, \quad a_2 = 1.5, \quad a_3 = -0.35355\ 33906.$

This polynomial is discussed in Chap. VII. As is pointed out there, there is much cancellation when it is evaluated for x near the value $x_0 = 0.707$. So, we now set

(3.16)
$$x = 0.707 + h$$

and rewrite the polynomial as a function of h.

For this, recall from Chap. II that, in the course of using Horner's method to evaluate this p at $x_0 = 0.707$, you obtain the numbers $b_0 = a_0$, and then $b_i = b_{i-1}x_0 + a_i$, $i = 1, 2, 3$. Then you have

$$\begin{aligned} p(x) &= (x - x_0)[b_0x^2 + b_1x + b_2] + b_3 \\ &= \qquad h\,q(x) \qquad\qquad + b_3, \end{aligned}$$

by Thm. 2.1 of Chap. II. There is, as you will see below, also cancellation in the evaluation of the quadratic polynomial q for x near $x_0 = 0.707$, so we go through Horner's method again, this time applied to q. This gives us $c_0 = b_0$, and then $c_i = c_{i-1}x_0 + b_i$, $i = 1, 2$. As before, you have

$$\begin{aligned} q(x) &= (x - x_0)\{c_0x + c_1\} + c_2 \\ &= \qquad h\,r(x) \qquad + c_2. \end{aligned}$$

There is, for our particular linear polynomial r here, cancellation in its evaluation near $x_0 = 0.707$, so, a final (and trivial) application of Horner's method gives us numbers $d_0 = c_0$ and $d_1 = d_0x_0 + c_1$ so that

$$\begin{aligned} r(x) &= (x - x_0)d_0 + d_1 \\ &= \quad h\,d_0 \quad + d_1. \end{aligned}$$

Now put all these calculations together and you get

(3.17)
$$p(x) = p(x_0 + h) = h[h\{h \cdot d_0 + d_1\} + c_2] + b_3.$$

The only requirement is that we carry out these calculations of the b_i's, c_i's and d_i's exactly so as not to lose accuracy. It is possible to make intelligent use of your calculator here, even though we are going to carry more than eight or ten digits in the calculations. See the discussion of double precision arithmetic at the end of Sect. 1. But there is nothing wrong with carrying out these few calculations by hand on a suitably large piece of paper. So, here goes:

3. Some ways to reduce cancellation

		$b_0 = a_0 = 1$
	0. 707	$b_0 x_0$
	-2. 1213 20344	$+ \; a_1$
	-1. 4143 20344	$= \; b_1$
0. 707 ×		
	9 9002 42408	
	990 02424 08	
	-0. 99992 44832 08	$b_1 x_0$
	1. 5	$+ \; a_2$
	0. 50007 55167 92	$= \; b_2$
0. 707 ×		
	3 50052 86175 44	
	3500 52861 7544	
	0. 3 53553 39037 1944	$b_2 x_0$
	-0. 3 53553 3906	$+ \; a_3$
	-0. 0 00000 00022 8056	$= \; b_3$

Notice the cancelling of nine digits in this calculation of $p(0.707) = b_3 = -0.22805\ 6 \times 10^{-9}$.

We continue.

		$c_0 = b_0 = 1$
	0. 707	$c_0 x_0$
	-1. 4143 20344	$+ \; b_1$
	-0. 7073 20344	$= \; c_1$
0. 707 ×		
	4 9512 42408	
	495 12424 08	
	-0. 50007 54832 08	$c_1 x_0$
	0. 50007 55167 92	$+ \; b_2$
	0. 00000 00335 84	$= \; c_2$

Notice the cancelling of seven digits in this calculation of $q(0.707) = c_2 = 0.33584 \times 10^{-7}$. Finally,

$$\begin{array}{rr} & \\ 0.707 & \\ -0.7073 & 20344 \\ \hline -0.0003 & 20344 \end{array} \qquad \begin{array}{rl} & d_0 = c_0 = 1 \\ & d_0 x_0 \\ + & c_1 \\ \hline = & d_1 \end{array}$$

Even in the calculation of d_1, there is cancelling of <u>three</u> digits.

The upshot of this calculation is that the cubic given by (3. 15a) and (3. 15b) can also be written as

$$p(0.707 + h) = h[h\{h - 0.320344 \times 10^{-3}\} + 0.33584 \times 10^{-7}] - 0.22805\ 6 \times 10^{-9} . \tag{3. 18}$$

For example, evaluation of (3. 15) at the point $x = 0.70781\ 25$ on the HP-33E using Horner's method gives the value 0 exactly. By contrast, using instead (3. 18), with $h = 0.00081\ 25$, on the same calculator gives the fairly accurate value

$$p(0.70781\ 25) \cong 1.2413\ 08594 \times 10^{-10} .$$

<u>Problem 3. 9.</u> Use Horner's method just once to rewrite the polynomial given by (2. 1) accurately into the form

$$\begin{aligned} p(x) &= (x - x_0)(b_0 x^3 + b_1 x^2 + b_2 x + b_3) + b_4 \\ &= \quad h\, q(x) \quad + b_4 \end{aligned}$$

with $x_0 = 4.45$. Then use this new form to calculate $p(40/9)$. Compare the value you get with the incorrect value

$$0.00060\ 96000$$

obtained in Sect. 2 and the correct value

$$4/9^4 = 0.00060\ 96631\ 61103\ldots .$$

<u>Problem 3. 10.</u> For $x = y = 0.7071$, there is severe cancellation in calculating the expressions

$$x^2 + y^2 - 1 \tag{3. 19}$$

$$2xy - 1 . \tag{3. 20}$$

Put $x = 0.7071 + h$ and $y = 0.7071 + k$, and get expressions in h and k for which the cancellation is considerably less severe.

Chapter V

NUMERICAL DIFFERENTIATION

0. Guide for the reader.

Read this chapter toward the end of the instruction on differentiation. However, pay special attention to the first four sentences of Sect. 1. Then read Sect. 3. For this you will have to skim through Sections 1 and 2 enough to learn what the booby traps are, and what the key formulas are. This should provide all that you need from this chapter for your calculus course.

If you plan to go on in engineering, physics, chemistry, or such, you should not skim too lightly over Sections 1 and 2. Now is the best time to get some help from your instructor or teaching assistant if you should encounter a difficulty.

1. Danger of numerical inaccuracy.

Very commonly, a function is defined in terms of some formula. The rules of calculus will usually enable one to write out a formula for the derivative. If one wishes the numerical value of the derivative at some point, it is STRONGLY recommended that one use the calculator to evaluate that formula for the derivative directly. This should give a fairly accurate value.

However, a function can be defined as the root of an equation, or in some other way, so that one cannot so easily write out a formula for the derivative. Presumably one can calculate the value of the function (at least approximately).

So we come down to the following question. How can we make use of function values to estimate a value for the derivative?

At first sight, this problem seems simple enough. Since

$$f'(x) = \lim_{h \to 0} \frac{f(x+h) - f(x)}{h} ,$$

we know that the difference quotient

(1.1) $$\frac{f(x+h) - f(x)}{h} ,$$

computable from function values alone, provides as good an approximation to $f'(x)$ as we could wish provided we choose h sufficiently small. But there is the problem: What is "sufficiently small"? We cannot take just any old h and expect the difference quotient (1.1) to be a good enough approximation to $f'(x)$. We discussed this difficulty already in Chap. III. There we made the point that "too large" an h makes (1.1) a poor approximation to $f'(x)$, while a theoretically "sufficiently small" h may make it impossible to evaluate (1.1) accurately on a calculator.

For example, the polynomial

$$f(x) = x^{100}$$

has the derivative

$$f'(x) = 100x^{99} ,$$

and therefore $f'(1) = 100$. Yet, with $x = 1$ and $h = 0.1$, we get

$$\frac{f(x+h) - f(x)}{h} \cong \frac{13780.61234 - 1}{0.1} = 1\,37796.1234 .$$

Since we know the exact value of $f'(1)$ in this case, we know that this number has little to do with $f'(1)$. So, we simply use smaller values of h. Some results are shown in Table 1.1, which was calculated on an HP-33E (many of the zeros shown are just copied from the display, and are not significant).

Table 1.1

h	calculated difference quotient
10^{-1}	1 37796.1234
10^{-2}	170.48138 29
10^{-3}	105.11569 80
10^{-4}	100.49662 00
10^{-5}	100.04950 00
10^{-6}	100.00500 00
10^{-7}	100.00000 00
10^{-8}	100.00000 00
10^{-9}	100.00000 00

We have carefully chosen the values of h so as to mislead the reader into jumping to the conclusion that Table 1. 1 shows that as one decreases h the calculated value of the difference quotient approaches f'(1), and indeed is exactly equal to it for $h \leq 10^{-7}$. However, let us try some other values of h. The results, as calculated on the HP-33E, are shown in Table 1. 2.

Table 1. 2

h	calculated difference quotient
$1.2345\ 49999 \times 10^{-5}$	100. 05710 59
$1.2344\ 99999 \times 10^{-6}$	99. 96597 821
$1.2349\ 99999 \times 10^{-7}$	99. 59514 178
$1.2499\ 99999 \times 10^{-8}$	96. 00000 008
$1.4999\ 99999 \times 10^{-9}$	66. 66666 671
$4.9999\ 99999 \times 10^{-10}$	0. 00000 0000

The difference quotient (1. 1) should get closer to f'(1) as h decreases. However, as shown in Table 1. 2, the calculated approximation to the difference quotient does not do so. At first, for $h = 10^{-1}$, it is extravagantly large. But then it begins to decrease toward f'(1). It behaves beautifully for the h's of Table 1. 1. However, for the h's of Table 1. 2, the first two entries get closer, but subsequent entries get worse, finally going to the entirely irrelevant value zero.

What is happening is that as h gets quite small, the two terms in the numerator of (1. 1) get very close together, so close that when we subtract them most of the initial digits cancel out, leaving a calculated difference that does not have too much relationship to what the true difference should be. Recall that the HP-33E calculator used for calculating the tables can hold at most 10 digits. When h gets small enough, the 10 digits of f(x + h) that the calculator can hold will be the same as the 10 digits of f(x) that the calculator can hold, and the calculator will have to give zero for f(x + h) - f(x), and hence zero for (1. 1). Thus for x = 1 and $h = 4.9999\ 99999 \times 10^{-10}$ in Table 1.2, the value of x + h would show on the calculator as 1. 0000 00000, the same as x. So of course the calculator will give identical values for f(x + h) and f(x). And, as we saw, even before h gets that small, the calculator will give very few correct digits for f(x + h) - f(x) because of cancellation of many of the initial digits (see Sect. 2 of Chap. IV)

This is further illustrated by the fact that different calculators will give very different results for the entries of Table 1. 2. See the related Table III. 2. 1, where quite divergent results from three different calculators are cited. As h becomes smaller, these entries are more and more contaminated by roundoff , and roundoff differs from calculator to calculator. Therefore not only will a calculator give, for certain small values of h, an answer which is seemingly unrelated to the true value of (1. 1), but these answers will differ from calculator to calculator. Or course, all calculators will give the answer 0 when h is small enough, regardless of what f or x might be.

This is an illustration of the two horns of the dilemma when using the difference quotient (1. 1) to approximate a derivative: Too large a step h produces numbers unrelated to the value of the derivative while too small a step h produces the value 0, which is most likely also unrelated.

We could use a calculus trick to compute (1. 1) more accurately for the case where $f(x) = x^{100}$. Refer to Rule 2 in Sect. 2-2 of T-F. By the binomial theorem

$$f(x+h) = (x+h)^{100}$$

$$= x^{100} + 100x^{99}h + \frac{(100)(99)}{2}x^{98}h^2 + \frac{(100)(99)(98)}{6}x^{97}h^3$$

$$+ \frac{(100)(99)(98)(97)}{24}x^{96}h^4 + (\text{terms in } x \text{ and } h)\cdot h^5 .$$

Subtracting $f(x) = x^{100}$ gives

$$f(x+h) - f(x) = 100x^{99}h + \frac{(100)(99)}{2}x^{98}h^2 + \frac{(100)(99)(98)}{6}x^{97}h^3$$

$$+ \frac{(100)(99)(98)(97)}{24}x^{96}h^4 + (\text{terms in } x \text{ and } h)\cdot h^5 .$$

So

$$(1.\ 2) \qquad \frac{f(x+h) - f(x)}{h} = 100x^{99}\{1 + \frac{99}{2}\frac{h}{x} + \frac{(99)(98)}{6}(\frac{h}{x})^2$$

$$+ \frac{(99)(98)(97)}{24}(\frac{h}{x})^3 + (\text{terms in } x \text{ and } h)\cdot h^4\} .$$

On the right, if we neglect the terms involving h^4, we have a polynomial in h/x. Evaluating this by Horner's method (see Sect. 1 of Chap. II) we get for $x = 1$ and $h = 1.2345\ 49999 \times 10^{-5}$

$$\frac{f(x+h) - f(x)}{h} \cong 100.06113\ 49 ,$$

instead of the number given in Table 1. 2. Indeed, for $x = 1$ and $h = 10^{-5}$, we get

$$\frac{f(x+h) - f(x)}{h} \cong 100.04951\ 62 ,$$

so that the value in Table 1. 1 was appreciably in error.

For still quicker evaluation of the polynomial on the right of (1. 2), we write it as

$$1 + \frac{99}{2}(\frac{h}{x})(1 + \frac{98}{3}(\frac{h}{x})(1 + \frac{97}{4}(\frac{h}{x}))) .$$

In this particular example, we had available a calculus trick to obtain an accurate difference quotient. But note that this trick is exactly the same trick that lets us produce the formula for $f'(x)$ for this case. When we do not know a formula for $f'(x)$, we are even less likely to know such a trick for avoiding cancellation in calculating the difference quotient. In that situation, we find ourselves on the horns of the dilemma: Too large an h will produce an inaccurate approximation while too small an h will induce cancellation and, again an inaccurate approximation. We must somehow choose h appropriately in the middle.

Problem 1. 1. Find the "best" h of the form 10^{-n} for some (positive or negative) integer n to approximate $f'(x)$ by (1. 1) (and without using the calculus trick), in case

(a) $f(x) = (x/3)^{50} + 6, \quad x = 2$

(b) $f(x) = (x/10^{10} + 1)^{20}, \quad x = 1$

(c) $f(x) = (10^{10}x - 1)^2, \quad x = 1$.

Remark. The exponentiation key y^x would be appropriate for the various calculations in this section.

Note that in each part of Prob. 1. 1, it was very easy to get a quite accurate value of the derivative by carrying out the differentiation and then evaluating the formula for the derivative by the calculator. As we said at the beginning of the section, if you can get a formula for the derivative, then, by all means, use it.

2. Approximation of derivatives by difference quotients.

So far, we have considered using the difference quotient

$$\frac{f(a+h) - f(a)}{h} \tag{2. 1}$$

as an approximation to $f'(a)$. We now wish to argue that the so-called centered difference quotient

$$\frac{f(a+h) - f(a-h)}{2h} \tag{2. 2}$$

is often a better approximation to $f'(a)$. This appears to be so in a situation like the one depicted in Figure 2. 1.

In Fig. 2. 1, the derivative $f'(a)$ is the slope of AB. (2. 1) is the slope of AC. (The discrepancy in slopes is readily apparent.) (2. 2) is the slope of DC, which is certainly close to the slope of AB. We now try to explain why (2. 2) is usually a better

Figure 2. 1

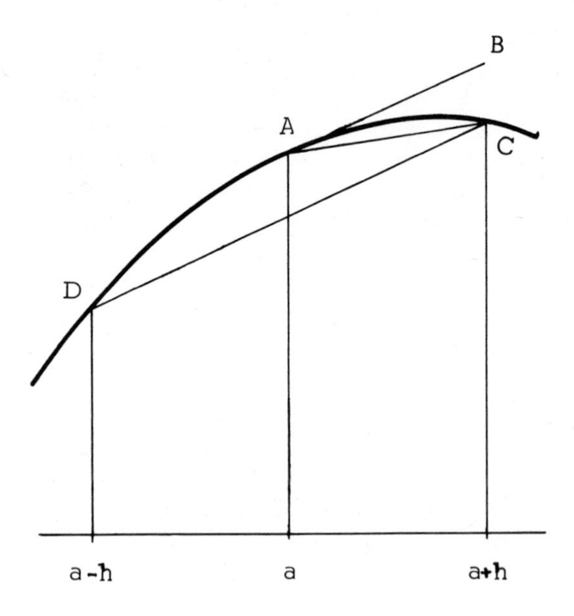

approximation to $f'(a)$ than (2. 1) is by looking at the errors in these approximations.

For this, we can use the Extended Mean Value Theorem (Special Case), as given in Sect. 3-10 of T-F. Taking $b = a + h$ there in Eq. (2) gives

$$f(a+h) = f(a) + f'(a)h + f''(\xi)h^2/2 ,$$

for some ξ between a and $a + h$.

If your text does not mention an Extended Mean Value Theorem, it may mention Taylor series with remainder, from which the result above can be derived. In any case, rest assured that it is true. In it, subtract $f(a)$ from both sides, and divide by h. This gives

$$\text{(2. 3)} \qquad \frac{f(a+h) - f(a)}{h} = f'(a) + \frac{1}{2}f''(\xi)h ,$$

for some ξ between a and $a + h$.

Next we consider (2. 2). By the Extended Mean Value Theorem (General Case), as given in Eq. (7) of Sect. 3-10 of T-F, but with $b = a + h$, we get

$$f(a+h) = f(a) + f'(a)h + \frac{1}{2}f''(a)h^2 + \frac{1}{6}f'''(\xi_1)h^3 ,$$

for some ξ_1 between a and $a + h$.

As above, if your text does not mention an Extended Mean Value Theorem, it may mention Taylor's series with remainder, from which the result above can be derived. So take our word for it. Replacing in the above h by $-h$, we get also

$$f(a-h) = f(a) - f'(a)h + \frac{1}{2}f''(a)h^2 - \frac{1}{6}f'''(\xi_2)h^3$$

for some ξ_2 between a and $a - h$. Now we subtract the second equation from the first and divide by $2h$ to get

$$\frac{f(a+h) - f(a-h)}{2h} = f'(a) + \frac{f'''(\xi_1) + f'''(\xi_2)}{12} h^2 \tag{2.4}$$

for some ξ_1 and ξ_2, both between $a - h$ and $a + h$.

Here we have h^2 in the final term on the right. If h is moderately small, h^2 will be very small. For example, if h is one thousandth, then h^2 is one millionth. So the left side of (2.4) should in general be much closer to $f'(a)$ than the left side of (2.3). And this has been accomplished without increasing the danger of cancellation in calculating the numerator!

In (2.4), the h^2 is multiplied by third derivatives of f. These could be unusually large, in which case the left side of (2.4) would not give a very good approximation for $f'(a)$. In our Fig. 2.1, we had $f''(x)$ reasonably constant, so that $f'''(x)$ is fairly close to zero. In such a case, (2.4) should be quite good, as is suggested by Fig. 2.1.

Figure 2.2

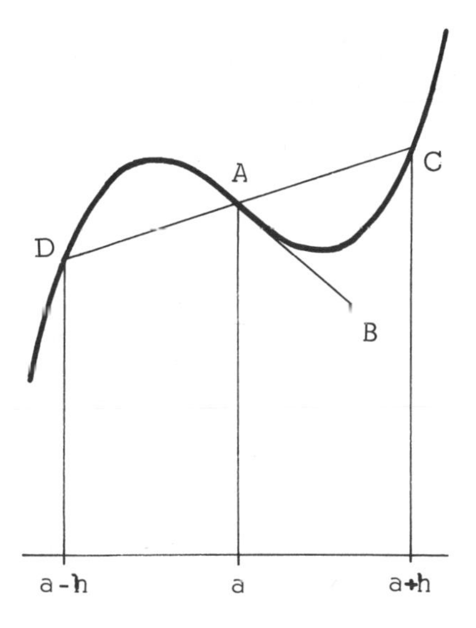

Let us look at a case where the third derivatives are large, as in Fig. 2. 2. Here the second derivative changes a lot in going from D to C, so that $|f'''(x)|$ must be large. The left side of (2. 3) is the slope of AC, and the left side of (2. 4) is the slope of DC. We purposely made them of equal slope. The derivative $f'(a)$ is the slope of AB, decidedly different from what either (2. 3) or (2. 4) would give.

<u>Problem 2. 1.</u> Show that for the function $f(x) = x^3 - x$ and $a = 0$, the error terms in (2. 3) and (2. 4) have the same value for all h.

<u>Remark.</u> Here the difficulty is not that we have large third derivatives, but that $f''(0) = 0$. So, with ξ close to 0, $f''(\xi)$ is small, so that (2. 3) gives an especially good approximation for $f'(a)$.

Large third derivatives or zero second derivatives are not common, so that usually (2. 4) is to be preferred to (2. 3). However, one should not put too much trust in (2. 4) unless one has some sort of guess as to the size of the third derivative.

Let us try a numerical example. Let $f(x) = e^x$. In Table 2 on p. A-21 of T-F is a short table of $f(x) = e^x$, and on p. A-22 is a short table of $f(n) = \ln n$. We reproduce a few entries in our Table 2. 1.

Table 2. 1

x	e^x	n	$\ln n$
1. 2	3. 3201	0. 4	-0. 9163
1. 3	3. 6693	0. 5	-0. 6931
1. 4	4. 0552	0. 6	-0. 5108
1. 5	4. 4817	0. 7	-0. 3567
1. 6	4. 9530	0. 8	-0. 2231

For $f(x) = e^x$, we have $f'(x) = e^x$, so that at $x = 1.4$ we have

$$f'(1.4) = e^{1.4} \cong 4.0552 . \tag{2. 5}$$

Now take $a = 1.4$ and $h = 0.1$. Then the left side of (2. 3) gives the approximation

$$f'(1.4) \cong \frac{4.4817 - 4.0552}{1.5 - 1.4} = 4.265 . \tag{2. 6}$$

The left side of (2. 4) gives the approximation

$$f'(1.4) \cong \frac{4.4817 - 3.6693}{1.5 - 1.3} = 4.062 . \tag{2. 7}$$

The improvement over (2. 6) is striking. Here the third derivative is not large and the second derivative is not zero.

Note that already we have our points close enough together that we have lost one significant digit by cancellation.

Problem 2. 2. If $f(n) = \ln n$, then $f'(n) = n^{-1}$. So $f'(0.6) \cong 1.6667$. Using the Table 2. 1, estimate $f'(0.6)$ by (2. 3) and (2. 4).

Problem 2. 3. Do Prob. 11 at the end of Sect. 1-9 of T- F, namely:

> 11. The following data give the coordinate s of a moving body for various values of t. Plot s vs. t on coordinate paper and sketch a smooth curve through the given points. Assuming that this smooth curve represents the motion of the body, estimate the velocity (a) at $t = 1.0$, (b) at $t = 2.5$, (c) at $t = 2.0$.

s (in ft.)	10	38	58	70	74	70	58	38	10
t (in sec.)	0	0. 5	1. 0	1. 5	2. 0	2. 5	3. 0	3. 5	4. 0

Problem 2. 4. In Prob. 2. 3, estimate the velocity: (a) at $t = 1.0$, (b) at $t = 2.5$, (c) at $t = 2.0$; use first (2. 3), and then (2. 4), taking $h = \Delta t = 0.5$ in each case.

Remark. It is shown in Prob. II. 1. 2 that the table above in Prob. 11 of T- F could have been derived from the equation:

$$s(t) = -16t^2 + 64t + 10 . \qquad (2.8)$$

Problem 2. 5. Assuming that (2. 8) is the correct formula for $s(t)$, explain why the formula (2. 4) gives the exact velocities.

3. Complicated expressions.

In some cases where one can get a formula for the derivative, as in

$$f(x) = \frac{x^3 + 5}{x^2} \sqrt{\frac{x^2 + 2}{x - 1}} , \qquad (3.1)$$

the expression is very complicated. To carry out the differentiation for this messy formula without making a mistake could be time consuming and tedious. Particularly if one needs only rough approximations for the derivatives, one might be well advised to use (2. 2). If one has a programmable calculator, the calculation would be fairly easy, since one has only to program the calculation of $f(x)$, after which two executions of the program, for the two input values of x, will give the needed two values of $f(x)$.

If it should happen, by some mischance, that one does need a very accurate value of the derivative of (3. 1), one had better go through the labor of differentiating it. However, it might be a wise precaution to get a rough approximate derivative by (2. 2) to check that one did not get a sign wrong, or an exponent in the wrong place, or something like that in carrying out the differentiation.

Problem 3. 1. For the f defined by (3. 1), derive the formula for $f'(x)$, and hence calculate $f'(2)$. Using $a = 2$, see what approximations are given by (2. 2) for $h = 10^{-1}$, $h = 10^{-2}$, and $h = 10^{-3}$. Do these approximations appear to confirm the error term in (2. 4), namely that dividing h by 10 should divide the error by approximately 100?

Actually, the differentiation of (3. 1) is not all that complicated if one uses logarithmic differentiation; see Sect. 6-9 of T- F, especially Example 4 therein.

Chapter VI

LOCAL APPROXIMATION BY A STRAIGHT LINE

0. Guide for the reader.

Calculus provides the tangent line as a local approximation to a function. By this we mean that the tangent line to f at a point x_1 describes well the behavior of f near that point x_1. In this chapter, we discuss the pros and cons to using the tangent as an approximation in numerical work. This chapter should therefore be read as soon as the derivative (or the differential) has been introduced in the calculus course. It would not hurt to review the first two sections of this chapter when you encounter in the calculus problems of such a sort as: "Approximate $\sqrt{9.001}$ by means of differentials."

In Sections 3 and 4, we use the secant as a local approximation. This provides the means to do linear interpolation, which is the same thing that you were taught earlier to call interpolation. You should find this a review, mainly devoted to how to do linear interpolation efficiently on a calculator.

These local straight line approximations to f are particularly useful when trying to determine an x for which

$$f(x) = 0,$$

as discussed in Chap. VII.

1. Approximation by the tangent.

The notion of derivative is built upon the following observation, true for many functions $y = f(x)$: As you look at the function in ever greater detail for x near a fixed point x_1, the function looks more and more like a straight line. In Figures 1.1, 1.2, and 1.3, we have shown this process for the function sin x, with $x_1 = 1$. When we concentrate on the interval from 0.9 to 1.1, a careful scrutiny is required to see the difference between the curve and a straight line; see Fig. 1.2. When we concentrate on the interval from 0.99 to 1.01, we would need a microscope to see the difference; see Fig. 1.3. (As we will see later, you will need many digits on your calculator to detect this minute difference.)

Figure 1. 1

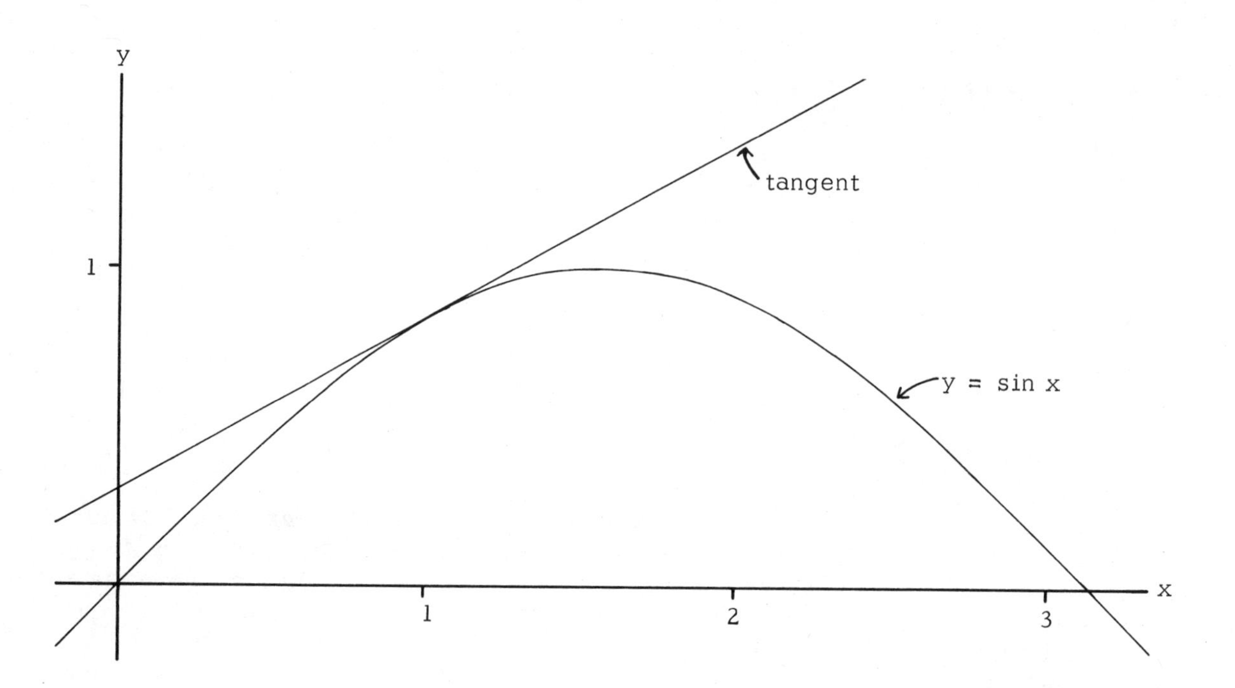

Figure 1. 2

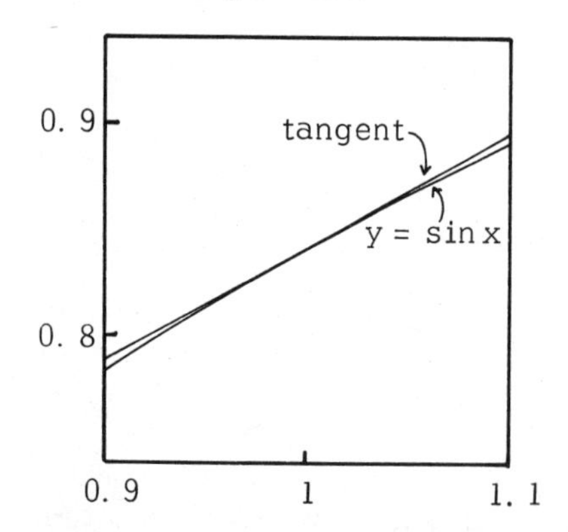

Figure 1. 3

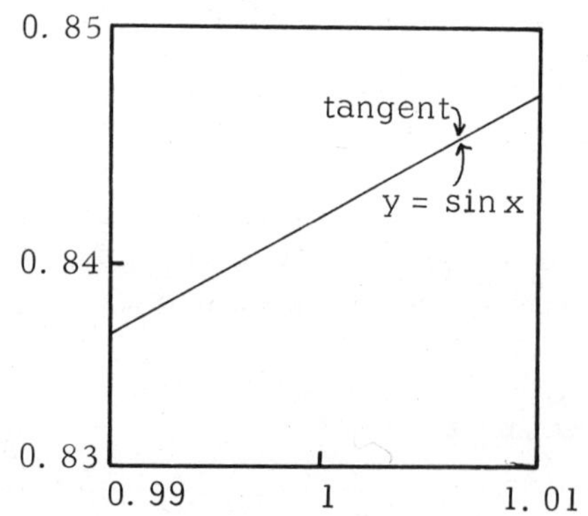

1. Approximation by the tangent

This straight line that f comes to resemble more and more as we look at ever smaller intervals containing x_1 is, of course, the tangent of f at x_1, i. e., the straight line

(1. 1) $$y = f(x_1) + f'(x_1)(x - x_1).$$

The fact that f comes to resemble this particular straight line more and more as we look at smaller and smaller intervals is expressed in mathematical terms as follows:

(1. 2) $$f(x) = f(x_1) + f'(x_1)(x - x_1) + \epsilon(x)(x - x_1)$$

$$\text{with} \lim_{x \to x_1} \epsilon(x) = 0.$$

In words, the difference between $f(x)$ and the tangent line

$$y = f(x_1) + f'(x_1)(x - x_1)$$

at x_1 goes to zero as $x \to x_1$, and does so faster than the difference $x - x_1$.

Incidentally, if your calculus book has not already told you this, then you should verify now that the function f is differentiable at x_1 if and only if for some a and b

(1. 3) $$f(x) = a + b(x - x_1) + \epsilon(x)(x - x_1)$$

with $\epsilon(x) \to 0$ as $x \to x_1$. If there is such an a and b, then the straight line $y = a + b(x - x_1)$ is necessarily the tangent, and a is necessarily $f(x_1)$, and b is necessarily $f'(x_1)$.

Having established the tangent line as a good local approximation to f near x_1, that is,

(1. 4) $$f(x_1 + \Delta x) \cong f(x_1) + f'(x_1)\Delta x$$

for "small" Δx, it seems reasonable to use it in this capacity, and calculus books usually make much of this idea. For this purpose, (1. 4) is often rewritten in the form

(1. 5) $$y + \Delta y \cong y + f'(x_1)\Delta x ,$$

and $f'(x_1)\Delta x$ is referred to as "the principal part of Δy."

Suppose one knows $f(x_1)$ (or can very easily calculate it). Should one use (1. 4) to estimate $f(x_1 + \Delta x)$? Before the advent of calculators, the answer was usually "Yes." However, let us stop and think a moment. One alternative is to go ahead and calculate $f(x_1 + \Delta x)$ outright. The other alternative is to calculate $f'(x_1)$, multiply it by Δx, and add to the known value of $f(x_1)$. Of course, this latter gives a poorer approximation for the answer, but the underlying assumption is that it is close enough. So it is basically a question of which is easier to calculate, $f(x_1 + \Delta x)$ or $f'(x_1)$. Without a calculator, $f'(x_1)$ was often easier to calculate. This is the reason for problems posed in calculus books such as the following, which are Problems 6 - 10 at the end of Sect. 2-6 of T-F:

> For each of the following functions $y = f(x)$, estimate $y + \Delta y = f(x_1 + \Delta x)$ by calculating y plus the principal part of Δy for the given data.
>
> 6. $y = \sqrt{x}$, $x_1 = 4$, $\Delta x = 0.5$
> 7. $y = \sqrt[3]{x}$, $x_1 = 8$, $\Delta x = -0.5$
> 8. $y = 1/x$, $x_1 = 2$, $\Delta x = 0.1$
> 9. $y = \sqrt{x^2 + 9}$, $x_1 = -4$, $\Delta x = -0.2$
> 10. $y = x/(1 + x)$, $x_1 = 1$, $\Delta x = 0.3$.

However, now you have a calculator. With it, you can often calculate $f(x_1 + \Delta x)$ straight off as easily as you can calculate $f'(x_1)$, perhaps more easily.

Problem 1.1. You will certainly have problems like the above in your calculus text. (If your calculus text is T-F, the above is exactly it.) Probably your instructor has assigned some of these, to be done the calculus way. This is important because it emphasizes the use of the tangent as a local approximation. By all means, carry out your assignment as given. But also calculate $f(x_1 + \Delta x)$ directly on your calculator for the same assigned problems, and verify that this takes less effort than the calculus way.

That's why you bought a calculator!

As we said, calculus texts tend to give quite a bit of stress to using the tangent line for a local approximation. Later in the section, T-F assign some more problems of this sort, to wit Problems 9-13 at the end of Sect. 2-13, namely:

> Use differentials to obtain reasonable approximations to the following:
>
> 9. $\sqrt{145}$ 10. $(2.1)^3$
>
> 11. $\sqrt[4]{17}$ 12. $\sqrt[3]{0.126}$
>
> 13. $(8.01)^{4/3} + (8.01)^2 - \dfrac{1}{\sqrt[3]{8.01}}$

And again, in the next chapter, T-F give another problem of this sort. See Prob. 7 at the end of Sect. 3-8, namely:

> 7. The Mean Value Theorem gives the equation
>
> $$f(b) = f(a) + (b - a)f'(c),$$
>
> c between a and b. When all terms on the right side of this equation are known, the equation determines $f(b)$ for us. Usually, however, $f'(c)$ is not known unless $f(b)$ is known. But when b is near a, then c will also be near

1. Approximation by the tangent

a, and the approximation

$$f'(c) \approx f'(a)$$

leads to the approximation

$$f(b) \approx f(a) + (b - a) f'(a).$$

Using this approximation, calculate

a) $\sqrt{10}$ by taking $f(x) = \sqrt{x}$, $a = 9$, $b = 10$;

b) $(2.003)^2$ by taking $f(x) = x^2$, $a = 2$, $b = 2.003$;

c) $1/99$ by taking $f(x) = 1/x$, $a = 100$, $b = 99$.

Problem 1. 2. If your calculus text gives another set of problems using the tangent line as a local approximation, proceed as in Prob. 1. 1.

The fact that, for problems in calculus texts that use the tangent line as a local approximation, your calculator gives $f(x_1 + \Delta x)$ directly more easily than by using (1. 4) or (1. 5) is no accident. The manufacturers of your calculator have read a calculus text or two, and such problems gave them the clue as to what functions should be built into the calculator. But this does not mean that the equations (1. 4) and (1. 5) are never useful. If it turns out that $f'(x_1)$ is much easier to calculate than $f(x_1 + \Delta x)$, then one of these equations could be quite handy. Such a case could be where

$$f(x) = \sqrt{\sin x + \cos x}.$$

Then

$$f'(x) = \frac{\cos x - \sin x}{2\sqrt{\sin x + \cos x}} = \frac{\cos x - \sin x}{2 f(x)};$$

If you save $\sin x_1$, $\cos x_1$, and $f(x_1)$ from the calculation of $f(x_1)$, then the calculation of $f'(x_1)$ is much easier than doing a fresh calculation of $f(x_1 + \Delta x)$. Another case arises when, for fixed x_1, you wish to calculate $f(x_1 + \Delta x)$ for several values of Δx. It is fairly quick to calculate $f'(x_1)$ and then use (1. 4) several times. If you do not have a programmable calculator, this is likely the better way to proceed. If you have a programmable calculator, you simply program it to calculate $f(x)$. Then, for each $x = x_1 + \Delta x$, you get $f(x)$ by inputting x and pressing the program key (keys).

Problem 1. 3. Calculate approximations for each of $f(0.1)$, $f(0.101)$, $f(0.102)$ and $f(0.103)$ if

$$f(x) = \sqrt{\frac{x^2 - 1}{x - 2}},$$

both by direct evaluation and by (1. 4), and compare efforts.

2. Approximation by an approximate tangent.

We discussed in Chap. V the possibility of approximating the derivative, for example by a difference quotient

$$f'(x_1) \cong \frac{f(a) - f(b)}{a - b} \tag{2. 1}$$

for a and b "near" x_1. Since the tangent provides only an approximation to f, there is the opportunity to replace $f'(x_1)$ in (1. 4) without appreciable further loss of accuracy by some approximation F to $f'(x_1)$, and so to use the approximation

$$f(x_1 + \Delta x) \cong f(x_1) + F\Delta x \ . \tag{2. 2}$$

This is attractive if the approximation F to $f'(x_1)$ is easier to obtain than the exact number $f'(x_1)$. For example, it looks much easier to evaluate the function

$$f(x) = \frac{x^3 + 5}{x^2 - 3} \sqrt{\frac{x^2 + 2}{x - 1}}$$

at two points and form the difference quotient, (2. 1), than to actually differentiate f and evaluate the derivative, particularly if you have a programmable calculator.

But, since we worried in Chap. V about the possible lack of accuracy in the approximation (2. 1), we must then deal with the following question: How good an approximation does F need to be in order for the right side of (2. 2) to remain a "good" approximation to $f(x_1 + \Delta x)$?

To answer this question, note first of all that if we make the choice of F in (2. 2) to be the particular difference quotient

$$\frac{f(x_1 + \Delta x) - f(x_1)}{\Delta x} \ , \tag{2. 3}$$

this would actually make (2. 2) into an <u>equality</u>. Therefore, our concern should really be how well F approximates (2. 3) rather than how well it approximates $f'(x_1)$! Second, we note that therefore the difference between $f(x_1 + \Delta x)$ and the right hand side of (2. 2) equals

$$\left(\frac{f(x_1 + \Delta x) - f(x_1)}{\Delta x} - F\right) \Delta x \ .$$

This has the consequence that, for "small" Δx, F need not be all that good an approximation to (2. 3).

To illustrate this, let $f(x) = x^{-1}$, and take $x_1 = 0.7$ and $\Delta x = 10^{-5}$. Then, by (III. 2. 3), the true value (to 10 digits) of (2. 3) is

$$-2.0407\ 87172.$$

However, from Table III. 2. 2, a calculated value came out to be

$$-2.0408\ 00000$$

(last five zeros not significant). As we are working to 10 digits, this is only a mediocre approximation. However, take this to be F in (2. 2). With $x_1 = 0.7$ and $\Delta x = 10^{-5}$, the right side of (2. 2) appears as

$$1.4285\ 71429\ -\ 0.0000\ 20408\ .$$

True enough, we have $F\Delta x$ correct to only 5 digits. However, these are to be added to the LAST 5 digits of $f(x_1)$. When the subtraction is performed, one gets

$$1.4285\ 51021\ ,$$

which is $f(x_1 + \Delta x)$ correct to 10 digits.

In other words, with Δx sufficiently small, $F\Delta x$ affects only the rightmost digits of the sum in (2. 2), and so needs to have only some of its leftmost digits correct.

Problem 2. 1. How accurate need the approximation F to (2. 3) be so that (2. 2) gives 10 digit accuracy, with f as in Problem 1. 3 and $x_1 = 0.1$ and

(a) $\Delta x = 0.003$ (b) $\Delta x = 0.0000\ 1$?

Problem 2. 2. Redo the calculations of Problem 1. 3 by evaluating f at two of the points, and then use the difference quotient at these points as the F in (2. 2) to obtain an approximation at the remaining points. Compare effort and accuracy with that for Problem 1. 3.

We made the point in Sect. 1 that it might be easier simply to evaluate the function f directly at $x_1 + \Delta x$ rather than to use (1. 4), particularly if one has a programmable calculator. The same remark applies to the use of (2. 2). But one should always keep in mind the following. Once one has gone to the trouble of calculating $f(x_1)$ and $f'(x_1)$ or F, the computational effort in using (1. 4) or (2. 2) is small, certainly smaller than for almost any function f not given by a single calculator key. It would then be cheaper (in the number of program steps required, say) to approximate f near x_1 by (1. 4) or (2. 2) than to record and repeat a program for the evaluation of f.

And let us not forget that sometimes there is no program for calculating f(x). So there is nothing to record and repeat in a programmable calculator. Refer to Chap. XV, which deals with the solution of differential equations. There we have given a formula for the derivative, and the problem is to recover the values of f(x). So getting an approximation for $f(x_1)$ may be an extended calculation which cannot be easily paralleled for $x_1 + \Delta x$. Certainly, there is no program for calculating $f(x_1)$ which one can just repeat to get $f(x_1 + \Delta x)$. As we have given a formula for the derivative at the start, this gives an approximation for F easily. Then, if Δx is sufficiently small, (2. 2) will be very useful.

<u>Problem 2. 3.</u> Suppose that by calculating from a differential equation we have found the approximations

$$f(1) \cong 0.12345\ 65432$$

$$f'(1) \cong 0.76543\ 23456.$$

Calculate approximations for $f(1 + \Delta x)$ for

(a) $\Delta x = 1.2345 \times 10^{-10}$ (b) $\Delta x = -1.2345 \times 10^{-9}$

(c) $\Delta x = 1.2345 \times 10^{-8}$ (d) $\Delta x = -1.2345 \times 10^{-7}$

(e) $\Delta x = 1.2345 \times 10^{-6}$.

<u>Remark</u>. You will note that if you start with Δx as in (a), then you get the other Δx's by successively multiplying by -10. So, as for Prob. III. 1. 6, you can program the changes of Δx on a programmable calculator. This saves inputting the other four Δx's.

3. Linear interpolation.

A commonly used choice for the approximate slope F in the linear approximation

$$f(x_1 + \Delta x) \cong f(x_1) + F\Delta x \ , \tag{3.1}$$

discussed in the preceding section, is the difference quotient

$$\frac{f(x_2) - f(x_1)}{x_2 - x_1}$$

with x_2 some point "near" x_1. This is the difference quotient (2. 3) for the specific choice $\Delta x = x_2 - x_1$. With this choice for Δx and F, (3. 1) actually becomes an equality, as we already pointed out in the preceding section. This means that the straight line

$$y = f(x_1) + \frac{f(x_2) - f(x_1)}{x_2 - x_1}(x - x_1) \tag{3.2}$$

agrees with the function f at the <u>two</u> points x_1 and x_2. We also say that the straight line <u>interpolates to</u> f <u>at the points</u> x_1 <u>and</u> x_2. In other words, it is a <u>secant</u> to (the graph of) f.

See Fig. 3. 1.

The straight line (3. 2) is a favorite approximating straight line, particularly in situations where f' is not available or is hard to get. It is well suited for getting

3. Linear interpolation

Figure 3. 1

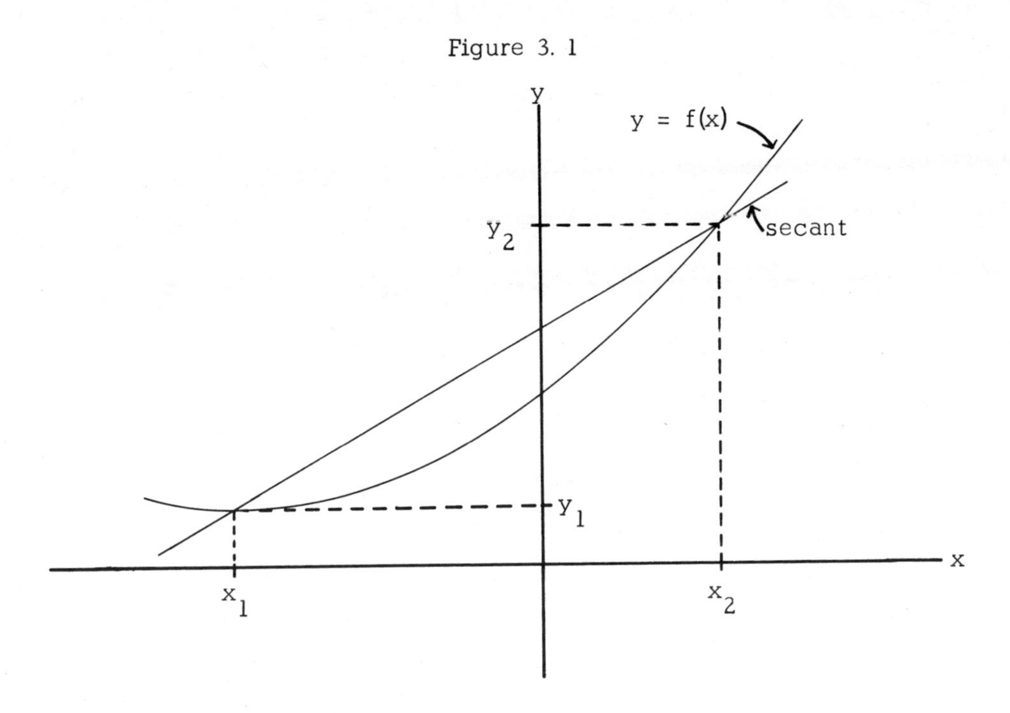

approximations for $f(x)$ when x lies between x_1 and x_2. Use of (3. 2) is called <u>linear interpolation</u>, because it involves the straight line (3. 2).

The formula (3. 2) is often written as

$$y = y_1 + \frac{y_2 - y_1}{x_2 - x_1}(x - x_1) . \tag{3. 3}$$

In this form, it should be familiar to the reader as the formula taught in high school for interpolation in a trig or log table, when the particular argument x is not listed in the table, but falls between two entries, x_1 and x_2, as in Table 3. 1.

Table 3. 1

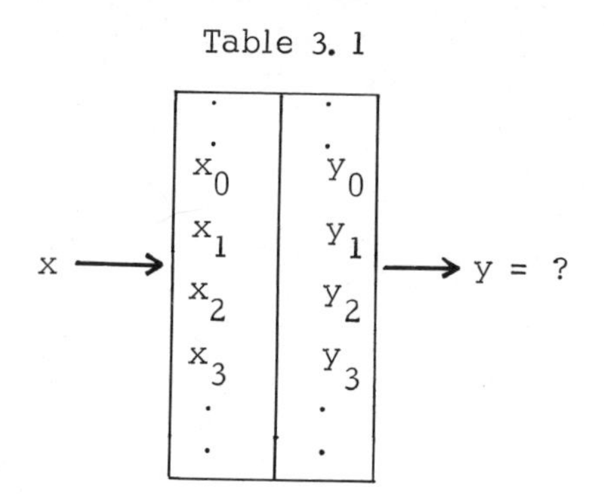

You can easily write a program to calculate the right side of (3. 3). However, x_1 and y_1 occur twice. If you have enough memory registers, you would store x_1 and y_1 when they are first input, for subsequent use. This is convenient, and reduces the ever present danger of error when inputting numbers.

Even the least expensive nonprogrammable calculator will usually have at least one storage register. Write

$$m = \frac{y_2 - y_1}{x_2 - x_1} .$$

Store this in the register. Now (3. 3) takes the form

$$y = y_1 + m(x - x_1) .$$

Then subtract x_1 from x, call m out of storage to multiply the difference, and add y_1.

If you have to use (3. 3) for more than one x with the same x_1, y_1, x_2, and y_2, this will save an inputting of each of these four quantities for each x after the first.

RPN

For an RPN calculator with three registers, you could use Program (3. 4).

(3. 4) Preparation:

y_2, [↑], y_1, [STO 2], [-],

x_2, [↑], x_1, [STO 1], [-], [÷], [STO 0],

Evaluation for a specific x:

x, [RCL 1], [-], [RCL 0], [×], [RCL 2], [+].

After execution of the Preparation, the slope $m = (y_2 - y_1)/(x_2 - x_1)$, the abscissa x_1, and the corresponding ordinate y_1 are stored respectively in registers 0, 1, 2. Hence we may now interpolate for as many x's as we please

RPN

RPN

between x_1 and x_2 by merely inputting x and running the Evaluation. On the other hand, if we merely wish to interpolate with a single x, then the two steps [STO 0] and [RCL 0] can be omitted. Then register 0 is not used.

It is, of course, possible to change the Program (3. 4) to apply to points (x_1, y_1) and (x_2, y_2) already in memory, say, x_1, y_1, x_2, y_2 stored respectively in registers 1, 2, 3, 4. Then you would replace the four steps y_2, [↑], y_1, [STO 2] by the two steps [RCL 4], [RCL 2], and the four steps x_2, [↑], x_1, [STO 1] by the two steps [RCL 3], [RCL 1]. This allows you to execute the Preparation without having to stop for inputs, after which the Evaluation runs for any x upon inputting x. This would be particularly suitable for a programmable calculator.

RPN

AE

For an AE calculator with three registers, you could use Program (3. 4)

(3. 4) Preparation:

y_2, [-], y_1, [STO 2], [=],

[÷],

[(], x_2, [-], x_1, [STO 1], [=], [STO 0],

Evaluation for a specific x:

x, [-], [RCL 1], [=], [×], [RCL 0], [+], [RCL 2], [=].

After execution of the Preparation, the slope $m = (y_2 - y_1)/(x_2 - x_1)$, the abscissa x_1, and the corresponding ordinate y_1 are stored respectively in registers 0, 1, 2. Hence we may now interpolate for as many x's as we please between x_1 and x_2 by merely inputting x and running the Evaluation. On the other hand, if we merely wish to interpolate with a single x, we could save a step and avoid using register 0 by replacing the [STO 0] at the end of the Preparation by [×] and using

[(], x, [-], [RCL 1], [)], [+], [RCL 2], [=]

for the Evaluation.

If we have x_1, y_1, x_2, y_2 already stored respectively in registers 1, 2, 3, 4, we would modify the Preparation part of Program (3. 4) by changing the inputs y_2, y_1, x_2, and x_1 respectively to [RCL 4], [RCL 2], [RCL 3], and [RCL 1]. We would also omit the two steps [STO 2] and [STO 1].

AE

Program (3. 4) can be adapted for use with a programmable calculator. However, there is no point in it unless one is going to use more than one x with a given set of x_1, y_1, x_2, y_2. Since the Preparation is run only once, there is no point in ever storing it. If (3. 4) is to be used for several x's with the same set of x_1, y_1, x_2, y_2, it would be good to store the Evaluation for reruns. Review Chap. 0, where there is a detailed discussion for a formula (0. 3), which is very similar to (3. 3).

There is ample opportunity to exercise linear interpolation in the next section.

We finish this section by recording an expression for the error in linear interpolation. It can be shown that, for a twice differentiable function f,

$$f(x) = f(x_1) + \frac{f(x_2) - f(x_1)}{x_2 - x_1}(x - x_1) + \frac{1}{2}f''(\xi_x)(x-x_1)(x-x_2) \tag{3.5}$$

for some ξ_x in the interval containing x, x_1, and x_2. Thus, for x between x_1 and x_2, the error in the approximation (3. 2) to $f(x)$ is bounded by the number $\frac{1}{8}M(x_1-x_2)^2$, with M a bound on the second derivative of f.

If, for example, $f(x) = \sin x$, and $x_2 - x_1 = h = .001$, then the error in linear interpolation would be no bigger than $\frac{1}{8}1(.001)^2 = .00000\ 0125$. This means that we would obtain accurate answers by linear interpolation in a six place table for sin if the argument spacing is no bigger than 10^{-3} (for radian arguments, of course).

Problem 3. 1. If one linearly interpolates in a log table for x between 2 and 3, and the arguments in the table are spaced .01 apart, how big an error could one possibly incur (given that the tabulated values are as accurate as required)?

4. Linear interpolation: examples.

Suppose we have a table of values for some function f. In it are given sundry values of x, opposite each of which is given an approximation for the unique $y = f(x)$. Inevitably, only a finite number of values for x are given, with their corresponding approximations for y. What do you do if you wish an approximation for $y = f(x)$ for a value of x which is not listed in the table?

Sometimes one is lucky, and it is very easy to calculate this $y = f(x)$ from the definition of f. However, situations will occur where this is not the case. It could be that the values in the table are the result of measurements from an experiment. There is no direct way to calculate an approximation for y, and it would be expensive to run the experiment over again, just to measure $y = f(x)$ for the particular x in question. Even if a formula for f is known, there could be reasons why it might be very difficult or time consuming to approximate y by attempting a direct calculation of $f(x)$.

One possible way to proceed is to graph the equation $y = f(x)$. The values in the table will give assorted points (x, y). You could then draw a smooth curve through these points, and try to read from the graph how high the curve is for the x in question. Don't laugh. This may well be the best thing to do. However, there are various other procedures, some of which might be better, particularly if a more accurate approximation for y is desired.

4. Linear interpolation: examples

One such scheme, which is widely used, is linear interpolation. One locates the arguments x_1 and x_2 in the table which are next above and next below the argument x, reads off the corresponding values y_1 and y_2, and then uses formula (3. 3) to obtain an approximation y to the sought-for value $f(x)$. See Table 3. 1.

We will now present several problems involving linear interpolation. In each problem, there will be only one x used with each set of x_1, y_1, x_2, y_2, so there is no point in storing Program (3. 4). Those with programmable calculators will have no advantage over those with nonprogrammable ones.

Problem 4. 1. Suppose we have Table 4. 1 for a function f.

Table 4. 1

x	f(x)
1	1
4	2
9	3

By linear interpolation, get approximations for $f(2)$ and $f(6)$.

Remark. The approximation for $f(2)$ is 1. 3333 33333. It is possible that the f for Table 4. 1 is $\sqrt{\ }$. If so, $f(2)$ would be

$$\sqrt{2} \cong 1.4142\ 13562\ .$$

This is not particularly close to 1. 3333 33333. In the first place, nobody has assured us that the f of Table 4. 1 really is $\sqrt{\ }$. It may be that 1. 3333 33333 is actually quite close to $f(2)$. In the second place, it was not claimed that linear interpolation would give a highly accurate estimate. What was claimed was that it is a procedure that requires only pressing a few keys to accomplish, and which can often give a reasonably good approximation. We gave a formula for the error in Sect. 3. There are fancier interpolation procedures that usually give better approximations than linear interpolation, but there was not room to discuss them.

Note. Did you use the calculator to get the estimates 1. 3333 33333 and 2. 4000 00000 for $f(2)$ and $f(6)$ respectively? You should have done these calculations in your head. After all, what is the slope of the line connecting (1 , 1) and (4 , 2)? According to Formula (1) in Sect. 1 - 4 of T - F, it is the rise divided by the run. But the rise, Δy, is 1 and the run, Δx, is 3. So the slope is 1/3. That is, y goes up 1/3 unit for every unit that x goes up. But x went up one unit, from 1 to 2. So y goes up 1/3 unit, from 1 to 4/3. Thus, the approximation for $f(2)$ is

$$\frac{4}{3}\ .$$

For $f(6)$, the rise is 1 and the run is 5. So y goes up $1/5$ unit for every unit that x goes up. But x went up two units, from 4 to 6. So y goes up $2/5$ units, from 2 to 2.4.

IMPORTANT. The calculator is a great help with hard problems. But we should remain alert to see when a problem can be done easily without using the calculator. Stay alert for these opportunities to beat the calculator. This will also help you in detecting when you have blundered, by pushing the wrong key or keys. Calculators are not for the mentally lazy.

Problem 4.2. For the reciprocal function, f, defined by

$$f(x) = \frac{1}{x}, \tag{4.1}$$

we have approximate values as given in Table 4.2.

Table 4.2

x	$f(x) = 1/x$
1.0	1.00000 00000
1.1	0.90909 09091
1.2	0.83333 33333
1.3	0.76923 07692

By linear interpolation, get an approximation for $f(1/0.9)$. Compare with an accurate answer.

Remark. We trust nobody evaluated $f(1/0.9)$ by putting

$$x = \frac{1}{0.9} \cong 1.1111\ 11111$$

in the display and pressing the reciprocal key. Surely, for the reciprocal function, f, it is obvious that $f(1/0.9) = 0.9$. Watch out about that mental laziness.

Problem 4.3. For sin we have approximate values as given in Table 4.3.

Table 4.3

x	$f(x) = \sin x$
0.4	0.38941 83423
0.5	0.47942 55386
0.6	0.56464 24734
0.7	0.64421 76872

By linear interpolation get an approximation for $\sin x$ for

$$x = \frac{\pi}{6} \cong 0.52359\ 87756 .$$

Compare with an accurate answer.

Remark. One can get an accurate answer for $\sin(\pi/6)$ by entering $\pi/6$ into the display and pressing the key (or keys) that cause the calculator to calculate sin (unless one forgot to set the calculator to operate in radians). But, we really had hoped that the reader would notice that

$$\frac{\pi}{6} \text{ radians} = 30^{\circ} ,$$

and would recall that

$$\sin 30^{\circ} = \frac{1}{2} .$$

The properly trained human brain can be a great labor saving device.

Problem 4.4. For $\sqrt{\ }$ we have approximate values as given in Table 4.4.

Table 4.4

x	$f(x) = \sqrt{x}$
0.0	0.00000 00000
0.1	0.31622 77660
0.2	0.44721 35955
0.3	0.54772 25575

By linear interpolation get an approximation for $\sqrt{0.04}$. Compare with an accurate answer.

Remark. After all we have said, surely nobody was so inattentive as to press the $\sqrt{x}$ key to get the exact answer

$$\sqrt{0.04} = 0.2 .$$

As for the linear interpolation itself, we have $x_1 = y_1 = 0$, so that (3.3) reduces to

$$y = \frac{y_2}{x_2} x = \frac{x}{x_2} y_2 = \frac{0.04}{0.1} (0.31622\ 77660) = (0.4)(0.31622\ 77660) .$$

You would still wish to do this final multiplication on the calculator, but surely this is much quicker than going through Program (3.4).

Moral. Keep your eye on the ball.

For curves of the form $y = f(x)$, going through (x_1, y_1) and (x_2, y_2), we have been seeking to approximate the number $y = f(x)$, for x between x_1 and x_2. We can turn the problem around and ask, for some y between y_1 and y_2, approximately what value of x will make $y = f(x)$ for that y. If we interchange x and y, this is exactly the same question we were asking earlier. So the answer is provided by interchanging x and y in (3.3), namely:

$$x = x_1 + \frac{x_2 - x_1}{y_2 - y_1}(y - y_1) . \tag{4.2}$$

To calculate this, we use exactly the same program as before, except that we now input (or store) the y's where formerly we inputted (or stored) the x's, and vice versa.

Problem 4.5. For the f of Table 4.1, find by linear interpolation the approximate values of x that will make $f(x) = 1.5$ and $f(x) = 2.5$.

Remark. If you picked up your calculator to work this problem, consider your knuckles rapped. Naturally the two required values of x would come halfway between 1 and 4 ($x = 2.5$) and halfway between 4 and 9 ($x = 6.5$).

Note. In case the f of Table 4.1 were $\sqrt{\ }$, then in Prob. 4.1 we were trying to evaluate $\sqrt{x}$ and in Prob. 4.5 we are trying to evaluate y^2. And in fact, 2.5 and 6.5 are not terribly bad approximations for $(1.5)^2$ and $(2.5)^2$.

Problem 4.6. For the f of Table 4.2, find by linear interpolation an approximate value of x that makes $f(x) = 0.8$. Compare with an accurate answer.

Remark. Surely there is no difficulty about getting an accurate answer. Put $f(x) = 0.8$ in (4.1) and solve for x.

Problem 4.7. For the f of Table 4.3, find by linear interpolation an approximate value of x that makes $\sin x = 0.6$. Compare with an accurate answer.

Remark. The accurate answer is $\sin^{-1}(0.6)$, which can be calculated by pressing suitable keys on the calculator.

Problem 4.8. For the f of Table 4.4, find by linear interpolation an approximate value of x that makes $f(x) = 0.4$. Compare with an accurate answer.

Note. Since the f of Table 4.4 was $\sqrt{\ }$, then in Prob. 4.4 we were trying to evaluate $\sqrt{x}$ and in Prob. 4.8 we are trying to evaluate y^2. So the exact answer is $(0.4)^2 = 0.16$.

Problem 4.9. For the function defined by

$$f(x) = x^4 - 4x^3 + 2x^2 - 4x + 1 , \tag{4.3}$$

Table 4.5 gives exact values. By linear interpolation, find an approximate value of x that makes $f(x) = 0$.

Table 4. 5

x	f(x)
0. 1	0. 6161
0. 2	0. 2496
0. 3	-0. 1199
0. 4	-0. 5104

Remark. An accurate answer is approximately

$$0.26794\ 91924\ .$$

Problem 4. 10. Consider Table 4. 6.

Table 4. 6

x	ln x + 2
3. 1	3. 1314 02111
3. 2	3. 1631 50810

We see that for $x = 3.1$, we have $\ln x + 2 > x$, while for $x = 3.2$, we have $\ln x + 2 < x$. So there must be an x between 3. 1 and 3. 2 at which the curves $y = x$ and $y = \ln x + 2$ cross each other. That is, there is an x such that $x = \ln x + 2$. By linear interpolation, find this x approximately.

Hint. Define the function f by

$$f(x) = x - \ln x - 2\ . \tag{4. 4}$$

From the values in Table 4. 6, we can find the values of $f(x)$ at $x = 3.1$ and $x = 3.2$. Now find an approximate value of x for which $f(x) = 0$.

Remark. An accurate value is approximately

$$3.1461\ 93220\ .$$

The process we have used in the last six problems, of using (4. 2) to estimate the x which would correspond to a specified y, is sometimes called inverse interpolation.

Chapter VII

ROOT FINDING

0. Guide for the reader.

Many problems in calculus involve finding a root of an equation. However, the topic is treated very sketchily, if at all, in most calculus books. This is because it involves numerical calculation, and practically every calculus book carefully avoids any sort of heavy arithmetic.

At certain points in the calculus, one absolutely has to find one or more roots for equations that arise during the solution of other problems. For example, to get maxima and minima, one has to solve

$$\frac{dy}{dx} = 0 .$$

Also, in curve sketching, one must again solve the equation above, to find where the slope of the tangent line changes from positive to negative, or vice versa (so the function changes from increasing to decreasing, or vice versa). Also, one gets intercepts by solving equations. If a curve is given implicitly as the locus of all points $A(x, y)$ in the plane which satisfy the equation $f(x, y) = 0$, one needs to solve this equation for x or y to find points on the curve. In integration, when one gets to the method of partial fractions, one must factor the polynomial in the denominator of the integrand, which requires getting roots. If, in addition, your calculus text touches on differential equations, you will again have to get roots of some equations there, called "characteristic equations."

How do calculus texts manage to avoid requiring the readers to do some numerical calculation? They give only a few problems, quite easy ones, which are carefully contrived so that the roots are obvious. Thus the texts are cleverly written so that you can get by without ever needing to learn the kind of material presented in this chapter.

But, if you go on in science or engineering you will encounter practical problems where you have to find roots of equations. These equations will not be carefully contrived to be easy. With a calculator, you can find the roots efficiently, by methods explained in this chapter. If you read it now, you can get help at any tricky points from your instructor or teaching assistant. You will also get a more realistic picture of the practical applications of calculus.

You might read the first four sections after you have become fairly familiar with differentiation. If your calculus text discusses Newton's method for finding roots, read Sections 5 and 6 when you get to that point. Read Sections 7 and 8 when you get to the relevant topic in your calculus course (see the titles of the sections). We stress the fact that Sect. 2 of Chap. VIII has some additional material on finding roots.

1. Zeros and sign changes.

Many problems of calculus involve finding the (or a) root of an equation. Suppose, for example, that we wish a root of the equation

$$x = \ln x + 2 . \tag{1.1}$$

By a root, we mean a value of x for which the equation is true.

Why would one think that the equation (1.1) has a root? It is usually advantageous to consider questions about roots in terms of zeros of functions. Define the function f by

$$f(x) = x - \ln x - 2 . \tag{1.2}$$

We say that c is a zero of a function f if and only if $f(c) = 0$.

It is quite clear that c is a zero of the function f of (1.2) if and only if it is a root of the equation (1.1).

So we are reduced to determining whether this f has any zeros. If so, how many are there, and where?

The function f is defined and continuous for positive x. This allows us to prove that f has a zero merely by exhibiting two positive numbers a and b for which $f(a)f(b) < 0$. For this then shows that $f(a)$ and $f(b)$ are of opposite sign, hence the number 0 lies between $f(a)$ and $f(b)$. Therefore, by the Intermediate Value Theorem for Continuous Functions, there must be a point c between a and b for which $f(c)=0$. (You find the Intermediate Value Theorem in T-F in Sect. 2-11, as Theorem 4.)

Now, by calculator

$$f(0.1) \cong 0.40259$$

$$f(1) = -1$$

$$f(4) \cong 0.61371 .$$

So there is a zero, c_1, of f between 0.1 and 1, and a zero, c_2, of f between 1 and 4. These are roots of the equation (1.1), of course.

If a function is not continuous, it need not have a zero when it changes sign. Consider the function, f, defined by

$$f(x) = \begin{cases} 1 & \text{if} \quad x \geq 0 \\ -1 & \text{if} \quad x < 0 . \end{cases}$$

It is negative for negative x and positive for positive x, but it does not have a zero as x goes from negative to positive.

Of course, we cannot tell by calculations whether a function is continuous. We need calculus for that. Calculus is also the key when we wish to find out how many zeros a function has.

Problem 1.1. Show that the function f of (1.2) has exactly two zeros, namely those we mentioned above, the c_1 between 0.1 and 1 and the c_2 between 1 and 4.

Hint. Obviously f has a zero when and only when the graph $y = f(x)$ crosses (or touches) the x-axis. So graph $y = f(x)$ for the f of (1.2). For a given x, you can get $\ln x$ by pressing one or two keys on your calculator. So you can quickly plot a number of points on the graph. If you have chosen them judiciously, they certainly suggest that the graph crosses the x-axis exactly twice. How can you be sure? Could the graph have some funny wiggles between or beyond the points you have plotted? This is where calculus comes to the rescue. Calculus texts explain carefully and thoroughly how you can settle such questions by seeing where the derivative of f is positive or negative; i.e., where f is increasing or decreasing. See Chap. 3 of T-F. You will learn after a while that

$$\frac{d}{dx} \ln x = \frac{1}{x} . \tag{1.3}$$

So it is easy to find $f'(x)$, for the f of (1.2). And it is easy to see where it is negative and where it is positive. So, with a few judiciously plotted points, you can readily establish that the graph $y = f(x)$ crosses the x-axis at exactly two points.

By calculus, you have proved conclusively that the function f of (1.2) has one and only one zero, c_1, between 0.1 and 1. What does the calculator say? On the HP-33E, let us calculate $f(x_i)$ for

$$x_1 = 0.15859\ 43396$$

$$x_2 = 0.15859\ 43397 .$$

For both these x's, and calculating $f(x)$ as $(x - \ln x) - 2$, the calculator gives 0, exactly.

Does that mean we were wrong, and there are indeed two zeros of f between 0.1 and 1? Not at all. Actually,

$$f(x_1) = -1.96\ldots \times 10^{-10}$$

$$f(x_2) = -7.26\ldots \times 10^{-10}$$

But the noise level in the calculated function values for f on the HP-33E (which is a 10 digit calculator) is about 10^{-9} for x near x_1, i.e., as great as, or greater than, the function values themselves in absolute value. So the computed function values have no significance. A further indication of this noise level can be obtained by calculating f(x) as x - (ln x + 2). Now, the HP-33E gives

$$f(x_1) = -4 \times 10^{-10}$$

$$f(x_2) = -3 \times 10^{-10}$$

instead of zero. We discuss this difficulty with noise in Sect. 4.

An accurate value of c is

$$c = 0.15859\ 43395\ 63039\ldots\ .$$

A different difficulty is illustrated by the function f given by

$$f(x) = \tan x - x\ . \tag{1.4}$$

Table 1.1 was calculated on the HP-33E, and shows that f(x) is negative for

$$x = 102.09\ 19663$$

and positive for

$$x = 102.09\ 19666\ .$$

Table 1.1

x	f(x)
102.09 19663	-0.0017 188
102.09 19664	-0.0006 766
102.09 19665	0.0003 657
102.09 19666	0.0014 080

So calculus assures us that f has a zero between those two values of x. But on a 10 digit calculator, such as the HP-33E, there are only two possible numbers between these two values of x, namely those shown in Table 1. 1; $f(x)$ is not particularly close to 0 for either of these.

In other words, the HP-33E would tell us that there is NO zero of $f(x)$ in the stated range. That is just a shortcoming of the calculator. In fact, $f(c) = 0$ for

$$c = 102.09\ 19664\ 64907 \ldots .$$

Even worse things can happen when you seek for zeros with a calculator, as you will see when we get to the discussion of noise in Sect. 4 below.

2. Getting a first approximation.

Root finding can be divided into two stages: (i) getting a first approximation; (ii) refining successive approximations. For the second stage, there are many methods available, and we discuss two of these in Sections 3 and 5. See also Sect. 2 of Chap. VIII. The first stage is much harder to deal with; unfortunately, there are really no fool-proof guidelines for finding that first approximation.

For example, what can you do to get a first approximation for a root of the equation

$$e^{\sin x} = \sin e^{x} \ ? \qquad (2.1)$$

As it happens, this equation has infinitely many roots. So, which root are we interested in? Usually it is (or should be) clear from the context roughly where to look for a root, or which root to look for. Once this is clear, then the best way to get a first approximation is to write the equation in the form

$$f(x) = 0$$

(recall how we got (1. 2) from (1. 1)), and then graph the function. This requires evaluating the function at a few equally spaced or judiciously chosen points. In this way, we may eventually discover a sign change in the function; if we have first verified the function to be continuous, this proves that there is a zero between the points where the function takes these opposite signs.

It is, of course, possible to graph such a function with quite fine spacing of the points, and produce thereby arbitrarily good approximations to a zero; by trial and error, so to speak. But, once one has a rough idea as to the whereabouts of a zero, it is usually more efficient to use one of the methods for refining an approximation discussed in Sections 3 and 5.

Problem 2. 1. Determine an interval which seems to contain the third positive root of equation (2. 1) and no other root of that equation.

Remark. You would have to analyze a very complicated derivative to be sure. Settle for doing a thorough job of plotting.

Problem 2. 2. Determine an interval which contains a zero of the polynomial p given by

$$p(x) = x^4 - 4x^3 + 2x^2 - 4x + 1 . \qquad (2.2)$$

Problem 2. 3. What is the total number of zeros of the p of (2. 2)?

Hint. In Sect. 3 of Chap. II there is a discussion of how to plot polynomials and their derivatives. If you use those ideas for the polynomial defined by (2. 2) and its derivative, you should get a graph of the p of (2. 2) that would settle the matter quite conclusively.

3. Bisection method.

To get started in finding a zero, you really should have located two points, $a < b$, at which the continuous function f has opposite signs. Then you are assured that there is a zero, c, of f somewhere between a and b. After than, we undertake to pin down the location of c better and better.

For an example, suppose that f is the polynomial p of (2. 2). In Fig. 3. 1, we

Figure 3. 1

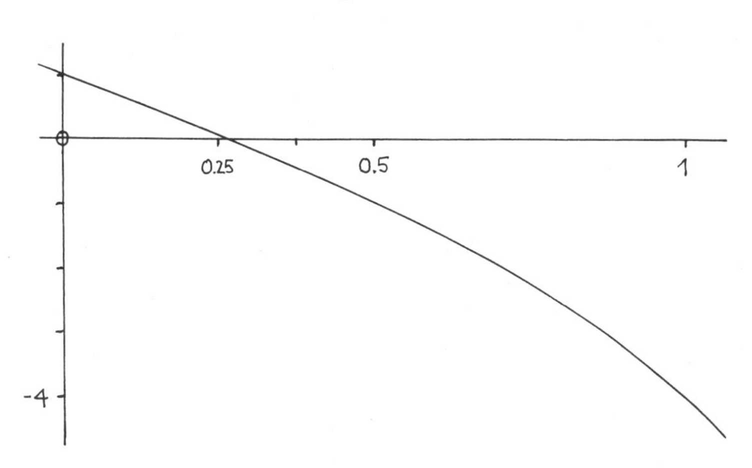

have drawn a bit of the graph. We find that $p(0) = 1$ and $p(1) = -4$. So we may take $a = 0$ and $b = 1$. As the graph clearly shows, there is a zero, c, between 0 and 1. See Fig. 3. 1.

The idea of the bisection method is to repeatedly reduce the length of the interval which is known to contain the zero c. Thereby, you pin down where c is. In fact we will slice off half of the interval at each step. That is why the procedure is called the bisection method.

Right now, we have the interval $(0, 1)$ that contains c. To get an interval half as long, we cut it in two. Of course that gives two intervals, $(0, 0.5)$ and $(0.5, 1)$. Which one contains c? We find that $p(0.5) = -0.9375$. So, at 0.5 and 1, p has the same sign. Not so good. But, therefore at 0 and 0.5, p has opposite signs. So there has to be a zero between 0 and 0.5. See Fig. 3.1.

So we have got down to an interval half as long as we started with. Well, cut it in half. This gives two shorter intervals, $(0, 0.25)$ and $(0.25, 5)$. Which one contains c? We find that

$$p(0.25) = 0.0664\ 0625\ .$$

At 0.25 and 0.5, p has opposite signs. So there has to be a zero between 0.25 and 0.5. See Fig. 3.1.

So now we cut the interval $(0.25, 0.5)$ into two halves, namely the intervals $(0.25, 0.375)$ and $(0.375, 0.5)$. We find that

$$p(0.375) \cong -0.40991\ 21090\ .$$

There never is any problem which of the shorter intervals to pick. At 0.375 and 0.5, p has the same sign, but therefore at 0.25 and 0.375, p has opposite signs. So we get down to the interval $(0.25, 0.375)$. See Fig. 3.1.

We could proceed, except that it is about time to get the calculator to do all this calculating for us. This requires us to describe carefully and in full detail what it is we have been doing.

We have been constructing a sequence of intervals (a_r, b_r), for $r = 0, 1, \ldots$. Specifically, for Fig. 3.1, we had

$$\begin{aligned}(a_0, b_0) &= (0, 1)\\(a_1, b_1) &= (0, 0.5)\\(a_2, b_2) &= (0.25, 0.5)\\(a_3, b_3) &= (0.25, 0.375)\ .\end{aligned}$$

Further, we always had

$$f(a_r) > 0 > f(b_r)\ .$$

This made certain that the interval (a_r, b_r) contains a zero of f. Now, to get from the interval (a_r, b_r) to the next smaller interval, we construct the midpoint

$$m_r = (a_r + b_r)/2$$

and consider the two intervals (a_r, m_r) and (m_r, b_r): We choose one or the other depending on the sign of the value of f at that midpoint m_r. For this, we calculate $f(m_r)$.

<u>Case 1</u>: $f(m_r) = 0$. Then we have actually found a zero and can stop.

<u>Case 2</u>: $f(m_r) > 0$. Then $f(m_r)$ and $f(b_r)$ are of opposite sign, hence the interval (m_r, b_r) is certain to contain a zero of f. So we take $a_{r+1} = m_r$ and $b_{r+1} = b_r$, and this insures that again

$$f(a_{r+1}) > 0 > f(b_{r+1}).$$

<u>Case 3</u>: $f(m_r) < 0$. Then $f(a_r)$ and $f(m_r)$ are of opposite sign, hence the interval (a_r, m_r) is certain to contain a zero of f. So, we take $a_{r+1} = a_r$ and $b_{r+1} = m_r$, and this insures that again

$$f(a_{r+1}) > 0 > f(b_{r+1}).$$

The whole process can be streamlined a bit as follows. Instead of carrying along both a_r and b_r, carry along the <u>midpoint</u> of the interval and its <u>half-length</u>, which are respectively

$$m_r = \frac{a_r + b_r}{2}, \qquad h_r = \frac{b_r - a_r}{2} .$$

These numbers are really more useful, as you will see. And of course you can always recover a_r and b_r from m_r and h_r:

(3. 1) $$a_r = m_r - h_r$$

(3. 2) $$b_r = m_r + h_r$$

Therefore, since $f(a_r) > 0 > f(b_r)$, we are assured that m_r is within a distance of h_r from a zero of f. Then our algorithm takes the form:

(0) Put $h_{r+1} = h_r/2$ and calculate $f(m_r)$;

(1) If $f(m_r) = 0$, STOP,

(2) If $f(m_r) > 0$, put

$$m_{r+1} = m_r + h_{r+1};$$

(3) If $f(m_r) < 0$, put

$$m_{r+1} = m_r - h_{r+1} .$$

If you are in doubt as to whether this algorithm really carries out the same process we described earlier, look at Figure 3. 2 where we have drawn a typical situation corresponding to case 2, or step (2) in the algorithm. We find $f(m_r) > 0$, so that the new interval is going to be the interval (m_r, b_r). Its length is half that of (a_r, b_r), so $h_{r+1} = h_r/2$ (which explains step (0)). Further, since m_r is an endpoint of the new interval, the midpoint m_{r+1} of the new interval is h_{r+1} away from it, or, $m_{r+1} = m_r + h_{r+1}$.

Figure 3. 2

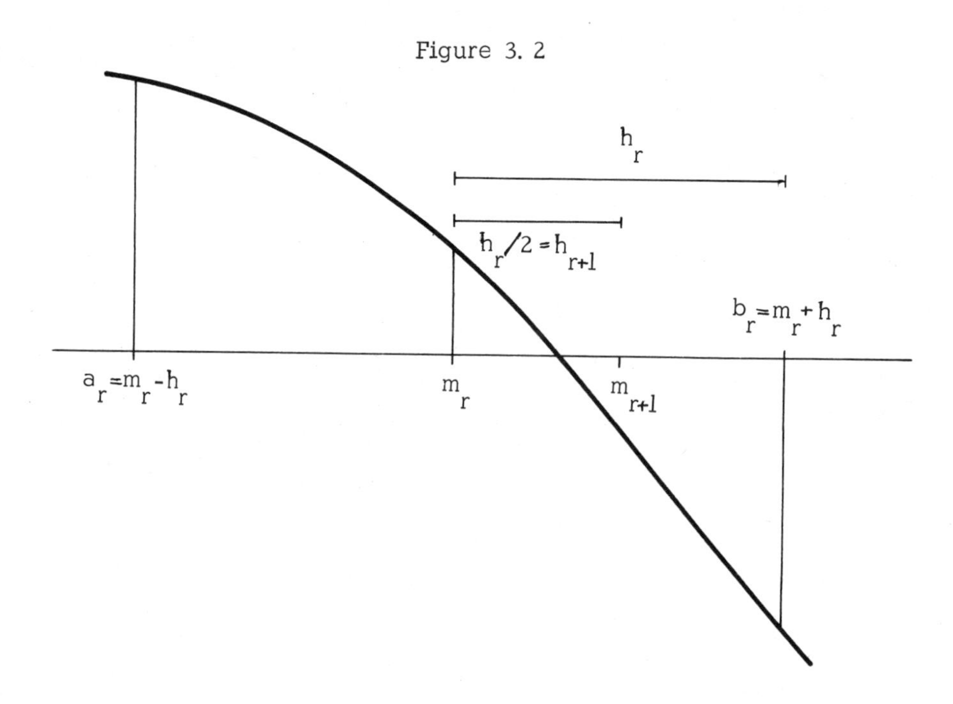

If you have a nonprogrammable calculator, you can watch the progress of the calculation and stop when it appears that you are as near to a zero as you wish (that is, h_r is as small as you would like). However, as step (0) involves the calculation of $f(m_r)$ for successive values of m_r, this could get to be a bit tedious, especially if the calculation of $f(x)$ is extensive.

We hope that you have a programmable calculator. If you do, you will find that it has keys that will determine if $f(m_r)$ is zero, positive, or negative. These keys actuate transfers to branches in a stored program. According to which case you have, the program will go to the branch where you have stored instructions appropriate to that case. So steps (1), (2), and (3) above can be handled in a stored program.

In the Program Appendix we give a specific storable program, called Program VII. 1, for the algorithm above. The program stops after step (3) with h_{r+1} in the display. If this is small enough, you take m_{r+1} as your approximation for the zero of f. Otherwise, you press a program key or two, and repeat steps (0) through (3).

As an example, the f defined by

(3. 3) $$f(x) = x^3 + x - 1$$

is everywhere defined and continuous. We have $f(0) = -1$ and $f(1) = 1$. Hence f has a zero between 0 and 1. So, in the preceding discussion, we take $a_0 = 0$ and $b_0 = 1$.

But, before you rush into applying the algorithm (or Program VII. 1) to this example, you should realize that something has changed. In the present example $f(a_0)$ is <u>negative</u>, whereas earlier we had f <u>positive</u> at the left endpoint of the interval known to contain a zero. If you check, you will find that our algorithm relied on this fact.

We could, of course, change the algorithm (or Program VII. 1) to fit this new situation. But, it is easier to fit the situation to the algorithm, as follows. Instead of looking for a zero of the function f given by (3. 3), we look for a zero of its negative, that is, of the function f given by

(3. 4) $$f(x) = -x^3 - x + 1 .$$

Clearly, any zero of the function given by (3. 3) is a zero of the function given by (3. 4) and vice versa. But now, for the function f given by (3. 4), we are back to the old situation, in which

$$f(a_0) > 0 > f(b_0) .$$

So, we are finally ready to use the bisection algorithm to find a zero of the function given by (3. 3), by finding instead a zero of its negative.

With $a_0 = 0$ and $b_0 = 1$, we have $h_0 = 0.5$ and $m_0 = 0.5$. Putting these into our program, we get the values given in Table 3. 1. These were actually calculated on the HP-33E, and transcribed from it. Needless to say, the subroutine to calculate $f(x)$ from x used Horner's method. If, for any value of r, we take the value m_r and mark off a distance h_r on each side of it, we have an interval in which we are assured there is a zero of f.

In Table 3. 1, the values given under $f(m_r)$ are given to the full accuracy that was obtainable on the HP-33E. Notice that most of them have fewer than 10 significant digits. Indeed, for $r = 15$, we have only 5 significant digits. What is happening is that as m_r approaches the zero of f, there is more and more cancellation in calculating $f(m_r)$. See Sect. 2 of Chap. IV for a discussion of this phenomenon.

The bisection method is sure to converge (in the absence of noise), but it does not converge particularly fast. After N iterations, it is guaranteed to reduce the length of the interval known to contain a zero by a factor of 2^N. In other words, $h_N = h_0/2^N$. Therefore, to reduce the initial error h_0 in the initial approximation m_0 by a factor of 10^{10} would take about $33 (\cong 10/\log_{10} 2)$ iterations. We have iterated 16 times in Table 3. 1, and do not quite have 5 digit accuracy.

Also, the method is occasionally inefficient, since it takes account of only the

Table 3. 1

r	h_r	m_r	$f(m_r)$
0	5×10^{-1}	0. 5	0. 375
1	2.5×10^{-1}	0. 75	-0. 17187 5
2	1.25×10^{-1}	0. 625	0. 13085 9375
3	6.25×10^{-2}	0. 6875	-0. 01245 11720
4	3.125×10^{-2}	0. 65625	0. 06112 67087
5	1.5625×10^{-2}	0. 67187 5	0. 02482 98642
6	7.8125×10^{-3}	0. 67968 75	0. 00631 38006
7	$3.9062\ 5 \times 10^{-3}$	0. 68359 375	-0. 00303 73930
8	$1.9531\ 25 \times 10^{-3}$	0. 68164 0625	0. 00164 60044
9	$9.7656\ 25 \times 10^{-4}$	0. 68261 71875	-0. 00069 37410
10	$4.8828\ 125 \times 10^{-4}$	0. 68212 89063	0. 00047 66193
11	$2.4414\ 0625 \times 10^{-4}$	0. 68237 30469	-0. 00010 84390
12	$1.2207\ 03125 \times 10^{-4}$	0. 68225 09766	0. 00018 41208
13	$6.1035\ 15625 \times 10^{-5}$	0. 68231 20118	0. 00003 78488
14	$3.0517\ 57813 \times 10^{-5}$	0. 68234 25294	-0. 00003 52930
15	$1.5258\ 78907 \times 10^{-5}$	0. 68232 72706	0. 00000 12781
16	$7.6293\ 94535 \times 10^{-6}$	0. 68233 49000	-0. 00001 70080

sign of the function values. For example, $|f(m_{15})|$ is less than one 27-th of $|f(m_{14})|$, so that the zero c should be much, much closer to m_{15} than to m_{14}. However, for the next approximation m_{16}, we simply average m_{14} and m_{15}, getting a result not nearly as good as m_{15}. In fact, we would have to carry on to m_{19} before we get finally an approximation better than m_{15}.

Algorithms that usually approach the zero more rapidly are known. One is coming up in Section 5.

Problem 3. 1. For the f of (1. 2), namely

$$f(x) = x - \ln x - 2 ,$$

find the zero between 0. 1 and 1 to about 5 digit accuracy.

Problem 3. 2. For the f of (2. 2), namely

$$f(x) = x^4 - 4x^3 + 2x^2 - 4x + 1 \ ,$$

find the zero between 0 and 1 to about 5 digit accuracy.

<u>Remark.</u> If one knows what accuracy one wishes to attain, as in the two problems just above, one can change the programs slightly. At present, in Program VII. 1, the program is not repeated unless the operator specifically calls for it. Instead, one could compare h_r with a preassigned tolerance; repeat the program if the tolerance is not attained and otherwise read m_r into the display and terminate.

4. The trouble caused by noise.

We discussed in Sect. 1 of Chap. IV the noise level associated with calculated function values. As we stated it there,

$$|\text{calculated value of } f \text{ at } x - f(x)| \leq \text{noise level.}$$

Ordinarily, the noise level is of little concern since it usually contaminates only the last one or two digits of the calculated function value. But, when x is near a zero of f, then $|f(x)|$ may well drop below the noise level and then the calculated value for $f(x)$ has no significance. We gave an example of this in Sect. 2. You can visualize this as in Fig. 4. 1. There we have drawn the graph of some function f. In addition, we

Figure 4. 1

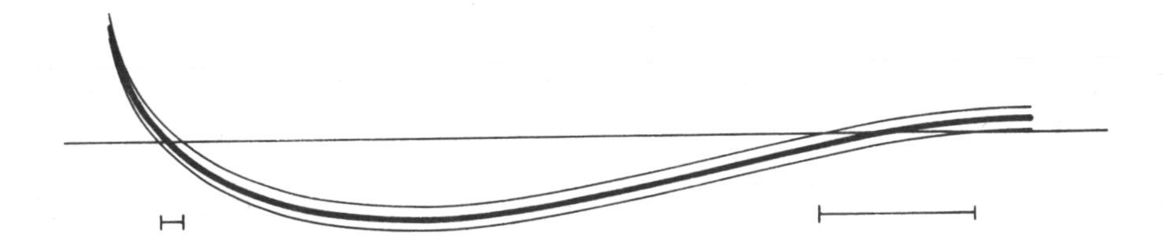

have drawn the functions f + noise level and f - noise level. The latter two functions form a band which encloses the function f and within which any calculated function value for f is bound to fall. But, - and that is the point of the noise level notion, - the calculated values may fall <u>anywhere</u> within that band.

The implication of this figure for the bisection method (or any other zero finding methods) is clear. Very near to a zero of f, the calculated function values are mostly noise and cannot be relied upon even to have the correct sign. In such a case, the bisection method would be misdirected. Further, the length of the interval in which noise is the dominant feature of the calculated function value is approximately

$$2 \times \frac{\text{noise level}}{|f'(x)|} .$$

Thus, this <u>interval of uncertainty</u> may be quite large in case $|f'(x)|$ is small near a zero of f. In Fig. 4. 1, for example, the situation at the right hand zero is much more dismal than at the left hand zero.

Here is a particularly nasty example of the trouble caused by noise.

Consider the f defined by:

(4. 1)
$$f(x) = x^3 - 2.1213\ 20344x^2 + 1.5x - 0.35355\ 33906 .$$

We calculate

$$f(0) = -0.35355\ 33906$$

$$f(1) = 0.02512\ 62654 .$$

So f has a zero between 0 and 1. Let us undertake to find it by the bisection algorithm. Remember to change the sign of the function before starting the algorithm. The results, using an HP-33E, are shown in Table 4. 1.

So $f(m_r) = 0$ when $m_r = 0.70779\ 60968$. Hence the latter must be a zero of f. NOT SO! The zero of f is $0.70771\ 70535$, which is not even particularly close to m_r.

In point of fact, from $r = 10$ onward, the true value of $f(m_r)$ is so near to 0 that it is completely masked by the roundoff error. The corresponding values listed in the column under $f(m_r)$ are noise, pure and simple. As they jump around, some negative and some positive, one of them accidentally hit 0, at $r = 18$. Just a fluke.

As it happens, we can show that for the f of (4. 1), we have EXACTLY

(4. 2)
$$\begin{aligned} f(x) = {} & (x - 0.707)^3 \\ & -3.2034\ 4 \times 10^{-4}(x - 0.707)^2 \\ & +3.3584 \times 10^{-8}(x - 0.707) \\ & -2.2805\ 6 \times 10^{-10} . \end{aligned}$$

Table 4. 1

r	h_r	m_r	$f(m_r)$
0	5×10^{-1}	0. 5	0. 00888 34766
1	2.5×10^{-1}	0. 75	-0. 00007 89159
2	1.25×10^{-1}	0. 625	0. 00055 35250
3	6.25×10^{-2}	0. 6875	0. 00000 75376
4	3.125×10^{-2}	0. 71875	-0. 00000 15780
5	1.5625×10^{-2}	0. 70312 5	0. 00000 00634
6	7.8125×10^{-3}	0. 71093 75	-0. 00000 00561
7	$3.9062\ 5 \times 10^{-3}$	0. 70703 125	0. 00000 00002
8	$1.9531\ 25 \times 10^{-3}$	0. 70898 4375	-0. 00000 00066
9	$9.7656\ 25 \times 10^{-4}$	0. 70800 78125	-0. 00000 00004
10	$4.8828\ 125 \times 10^{-4}$	0. 70751 95313	0. 00000 00006
11	$2.4414\ 0625 \times 10^{-4}$	0. 70776 36719	0. 00000 00002
12	$1.2207\ 03125 \times 10^{-4}$	0. 70788 57422	-0. 00000 00003
13	$6.1035\ 15625 \times 10^{-5}$	0. 70782 47070	-0. 00000 00002
14	$3.0517\ 57813 \times 10^{-5}$	0. 70779 41894	0. 00000 00001
15	$1.5258\ 78907 \times 10^{-5}$	0. 70780 94482	-0. 00000 00002
16	$7.6293\ 94535 \times 10^{-6}$	0. 70780 18188	-0. 00000 00001
17	$3.8146\ 97268 \times 10^{-6}$	0. 70779 80041	-0. 00000 00001
18	$1.9073\ 48634 \times 10^{-6}$	0. 70779 60968	0. 00000 00000

In Sect. 3 of Chap. IV ways to reduce cancellation are discussed. Amongst them is a way to discover formulas like (4. 2) and to verify that they are exactly correct. Indeed, (4. 2) is the same as (3. 18) in Chap. IV.

For x near 0. 7, the right side of (4. 2) is much less subject to cancellation than the right side of (4. 1). Correspondingly, the noise level in function values computed from (4. 2) is much lower. Using the right side of (4. 2), we find that at $r = 10$, $f(m_r)$ should be only about 1.5684×10^{-10} instead of 6×10^{-10}. At $r = 11$, $f(m_r)$ is negative, being about -5.6138×10^{-11} instead of 2×10^{-10}. A partial table of the bisection algorithm is given in Table 4. 2, using the much less noisy values of $f(m_r)$ obtained from (4. 2). Clearly m_r is progressing satisfactorily toward the true zero 0. 70771 70535.

For calculation on the HP-33E using (4. 1), the interval of uncertainty is roughly

Table 4. 2

r	h_r	m_r	$f(m_r)$
10	$4.8828\ 125 \times 10^{-4}$	0. 70751 95313	1.5684×10^{-10}
11	$2.4414\ 0625 \times 10^{-4}$	0. 70776 36719	-5.6138×10^{-11}
12	$1.2207\ 03125 \times 10^{-4}$	0. 70764 16016	7.4262×10^{-11}
13	$6.1035\ 15625 \times 10^{-5}$	0. 70770 26368	1.5721×10^{-11}
14	$3.0517\ 57813 \times 10^{-5}$	0. 70773 31544	-1.8458×10^{-11}
15	$1.5258\ 78907 \times 10^{-5}$	0. 70771 78956	-9.4160×10^{-13}
16	$7.6293\ 94535 \times 10^{-6}$	0. 70771 02662	7.4952×10^{-12}
17	$3.8146\ 97268 \times 10^{-6}$	0. 70771 40809	3.3033×10^{-12}
18	$1.9073\ 48634 \times 10^{-6}$	0. 70771 59882	1.1876×10^{-12}
19	$9.5367\ 43170 \times 10^{-7}$	0. 70771 69419	1.2465×10^{-13}
20	$4.7683\ 71585 \times 10^{-7}$	0. 70771 74187	-4.0800×10^{-13}

from 0. 707 to 0. 7085. Fortunately, the interval of uncertainty is very seldom anywhere near this bad.

Problem 4. 1. For your calculator, estimate the noise level near $x = 0.707$ in the function values for the function f given by (4. 1), as calculated from (4. 1). Then determine the interval of uncertainty about the zero of f near 0. 7077.

Problem 4. 2. For your calculator, estimate the noise level near $x = 0.707$ in the function values for the function f given by (4. 1), but as calculated from (4. 2). By how much is the noise level decreased when using (4. 2) instead of (4. 1)?

Problem 4. 3. For your calculator, estimate the noise level near $x = 0.2$ in the calculated function values for the function given by (1. 2). How accurately can you hope to determine the zero of f near 0. 2?

5. Zeros by Newton's method.

This topic is discussed in many calculus texts, for instance in Sect. 2-7 of T-F. The idea of the method is to use the tangent line instead of the function itself when looking for a zero. Fig. 5. 1 tells the whole story.

You have somehow gotten an approximation x_0 to the zero c of the function f. On evaluating f at x_0, you find that $f(x_0) \neq 0$. So you know that $x_0 \neq c$. Now, you don't know how to find a zero of f exactly, but you have no difficulty finding a zero of any particular straight line. So, in order to get closer to c, you determine the zero of the tangent to f at x_0. This tangent is the straight line given by

5. Zeros by Newton's method

Figure 5. 1

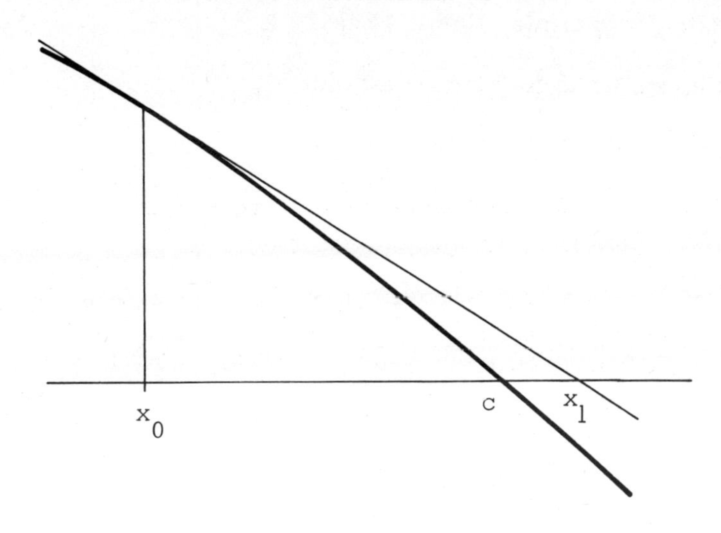

$$y = f(x_0) + f'(x_0)(x - x_0)$$

and it is very easy to find its zero x_1. You only have to solve

$$0 = f(x_0) + f'(x_0)(x_1 - x_0)$$

and this gives

$$x_1 = x_0 - \frac{f(x_0)}{f'(x_0)}, \tag{5.1}$$

(provided $f'(x_0) \neq 0$) If now x_0 is close enough to c so that the tangent is a good approximation to f near c, then you can count on x_1 to be an even better approximation for c than x_0 is. But then, you are certain to do even better by trying the whole process again with x_1 in place of x_0, getting

$$x_2 = x_1 - \frac{f(x_1)}{f'(x_1)}. \tag{5.2}$$

And away we go, generating a sequence $x_0, x_1, \ldots$ by the formula

$$x_{r+1} = x_r - \frac{f(x_r)}{f'(x_r)}. \tag{5.3}$$

This is called Newton's method. If x_0 is close enough to c, the successive x_r's will get closer and closer to c. In such a case, we say that the x_r's converge to c. The concept of convergence is discussed in the chapter of sequences and series in your calculus text: for example, Chap. 16 in T-F.

As an example, we try to find the larger root of

$$x = \ln x + 2 \tag{5.4}$$

by Newton's method. First we have to write the equation in the form

$$f(x) = 0 .$$

For this, we take

$$f(x) = x - \ln x - 2 . \tag{5.5}$$

Then by (1.3)

$$f'(x) = 1 - \frac{1}{x} . \tag{5.6}$$

A quick sketch shows the larger root to lie near 3, so we take $x_0 = 3$. Table 5.1 shows the successive iterates as calculated on a TI-30. We have only recorded the answers as shown in the display, i.e., we have not tricked the calculator into revealing all the digits of the answers carried internally. The TI-30 has but one memory register, and is AE with hierarchical arithmetic (remember My Dear Aunt Sally from Sect. 2 of Chap. 0). So, initially, we put $x_0 = 3$ into the display. After that, we repeated the following program.

[-], [(], [STO], [-], [ln x], [-], 2, [)], read off $f(x_r)$,

[÷], [(], 1, [-], [RCL], [1/x], [=], read off x_{r+1} .

The convergence is fast, as is typical for Newton's method if the initial guess x_0 is close enough to the zero.

Table 5.1

x_r	$f(x_r)$
3.	-.09861 229
3.1479 184	.00117 701
3.1461 934	.00000 015
3.1461 932	-1×10^{-10}
3.1461 932	0

5. Zeros by Newton's method

As we said in Sect. 2, root finding can be divided into two stages: (i) getting a first approximation; (ii) refining successive approximations. We have (5. 3) to take care of (ii). In calculus texts that discuss Newton's method, the problems assigned usually tell you what to take for your first approximation, x_0. That is, the author of the text has taken care of (i) for you. And, with the x_0 suggested by the author of the text, you get rapid convergence to a zero. How jolly.

In engineering and science courses, there is usually not anybody to help you out by suggesting an x_0. You are on your own.

There are some theorems available which guarantee convergence of Newton's method from certain starting values x_0. But these require you to know an awful lot about the function f. On the other hand, the proof of the pudding is in the eating. So, somehow locate a first guess x_0, for example by graphing f, and start Newton's method with it. If this leads to success, fine; if not, then you had better graph f with points more finely spaced in order to get a better guess. If you had a change of sign, you could instead use the bisection method for a few steps.

As an example, we now use Newton's method to look for the smaller root of the equation (5. 4),

$$x = \ln x + 2 .$$

We try $x_0 = 0.5$. Then $f(x_0) \cong -0.80685\ 28190$ and $f'(x_0) = -1$. Therefore, by (5. 1),

$$x_1 \cong 0.5 - (-0.80685\ 28190)/(-1) = -0.30685\ 28190 .$$

This is the end of the trail. We cannot form $f(x_1)$, since $\ln x$ is undefined for x negative, and hence (5. 2) cannot give us x_2. As we said, the matter of finding a suitable x_0 is sticky.

One thing you learn in an engineering or science course is not to give up if you don't make it on the first try. Let us experiment a bit. We calculate f(0. 4), f(0. 3), f(0. 2), and f(0. 1). From these, we pick $x_0 = 0.2$ as another possible starting point. As Table 5. 2 shows, this works excellently.

Table 5. 2

r	x_r	$f(x_r)$	$f'(x_r)$
0	0. 2	-0. 19056 20880	-4
1	0. 15235 94780	0. 03387 20410	-5. 5634 24955
2	0. 15844 78213	0. 00077 77640	-5. 3112 25940
3	0. 15859 42591	0. 00000 04270	-5. 3053 98478
4	0. 15859 43396	0. 00000 00000	-5. 3053 95278

Part of the folklore about the Newton method is that the number of correct digits in the approximation x_r doubles for each new step. Actually, this is not so in every case. The following application of calculus gives a more precise description of how the error in the Newton method iterates decreases from iteration to iteration.

By the Extended Mean Value Theorem (Special Case), which is Equation (2) of Sect. 3-10 of T-F, we have

$$f(c) = f(x_r) + f'(x_r)(c - x_r) + \frac{1}{2}f''(\xi)(c - x_r)^2 ,$$

where ξ lies between c and x_r. By (5. 3), we have

$$0 = f(x_r) + f'(x_r)(x_{r+1} - x_r) .$$

If we subtract this equation from the equation above, we get

$$f(c) = f'(x_r)(c - x_{r+1}) + \frac{1}{2}f''(\xi)(c - x_r)^2 .$$

But with c a root of $f(x) = 0$, we get from this

$$c - x_{r+1} = -\frac{f''(\xi)}{2f'(x_r)}(c - x_r)^2 . \tag{5. 7}$$

In case

$$\left| \frac{f''(\xi)}{2f'(x_r)} \right| \tag{5. 8}$$

is somewhere near unity, then indeed the number of correct digits in x_{r+1}, after the decimal place, would be very close to double the number of correct digits in x_r, <u>after</u> the decimal place. However, the number of correct digits, <u>after the decimal place</u>, could differ quite a bit from the number of correct significant digits. So, even with (5. 8) close to unity, we are not assured of doubling of the number of correct significant digits. And, of course, there is no particular reason why (5. 8) should be anywhere near unity.

If (5. 8) stays somewhere near constant as r increases, we will certainly have the rate of convergence increasing. This is what usually happens. As one can see from (5. 7), it takes really abnormal behavior on the part of (5. 8) to prevent an increase in the rate of convergence. Hence Newton's method usually gets the answer quite quickly.

How do we know we have the right answer? Table 5. 2 was calculated on an HP-33E. For $r = 4$, it says

$$f(0.15859\ 43396) = 0 , \tag{5. 9}$$

as plain as day. How far off might the calculator be? By the Mean Value Theorem, which is stated in Sect. 3-8 of T-F,

(5.10) $$f(x_4) - f(c) = f'(\xi)(x_4 - c) ,$$

for some ξ between x_4 and c. But for ξ between x_4 and c, we have $f'(\xi)$ very close to $f'(x_4)$. So

(5.11) $$f'(\xi) \cong -5.3054 ,$$

from Table 5.1. With the value of x_4 from Table 5.2 and the following accurate value of c

$$c = 0.15859\ 43395\ 63039 \ldots ,$$

we have

(5.12) $$x_4 - c \cong 3.6961 \times 10^{-11}$$

As $f(c) = 0$, we get from (5.10) a more accurate value than given by (5.9), namely

(5.13) $$f(0.15859\ 43396) \cong -1.961 \times 10^{-10} .$$

The difficulty is that we are encountering noise. This was discussed in detail in Sect. 4 of this chapter. The noise in the illustration given there was much worse than we are encountering here, but here it is bad enough that our calculator gave an approximation for $f(x_4)$ that was two units too high in the last digit.

Suppose we had had an accurate value for $f(x_4)$ in Table 5.2 instead of the erroneous one given there. The value of $f(x_4)/f'(x_4)$ would have been less than 5×10^{-11}; on a 10-digit calculator, adding this to x_4 would have produced no change. Then x_5 would have come out exactly the same as x_4. We did the best that is possible with a 10-digit calculator, in spite of the presence of a slight amount of noise.

This is a comforting conclusion. But we managed to reach it only because we somehow produced out of the blue the (first fifteen digits of the) exact value of c. We usually do not have this information available. So, how do we then know how well we did?

If we reach the point where $x_{r+1} = x_r$, then we have certainly done as well as was possible, starting from the particular first guess x_0 we used. Usually, x_r is then an accurate approximation for c. But it is not guaranteed to be that.

For example, take

(5.14) $$f(x) = x^{\alpha}$$

for $\alpha > 0$. Then (5.3) would give

(5.15) $$x_{r+1} = x_r - \frac{x_r}{\alpha} .$$

If $x_r = 1$ and $\alpha = 10^{12}$, then a 10-digit calculator will give $x_{r+1} = x_r$, leading one to think that $x_r = x_0$ is a root of

(5. 16) $$x^{\alpha} = 0 .$$

Some people write their program for Newton's method so that it displays only x_r. Then, when x_{r+1} agrees with x_r to the desired accuracy, they take x_{r+1} as the root. As this example shows, this is risky.

A simple way to check the accuracy of x_{r+1} as an approximation to a zero c of f is to evaluate f at a nearby point beyond x_{r+1}, for example at the point

$$b = x_{r+1} + (x_{r+1} - x_r).$$

If f has opposite signs at x_{r+1} and b, and f is continuous, then you can be sure that x_{r+1} is within $|x_{r+1} - x_r|$ of a zero c. Of course, you have to have enough of an idea about the noise level in the calculated function values to be certain that the function values at b and x_{r+1} are truly of opposite sign.

Problem 5. 1. Write a program to perform the calculations for Table 5. 2.

Problem 5. 2. Use a slight variant of the program written in Prob. 5. 1 to find the zero of

$$x = \ln x + 3$$

that is near $x = 5$, to ten significant digits, by Newton's Method.

Remark. This time, take $x_0 = 5$.

To get zeros for a polynomial by Newton's method is of course very easy. Given x_r, we calculate $p(x_r)$ and $p'(x_r)$, which is easy, and plug them into (5. 3). A few repetitions of this, and we have a good approximation for c. This would not be unduly laborious on a nonprogrammable calculator, but with a programmable calculator, it is extremely fast. Recall that Program II. 4 in the Program Appendix will do the following: after putting x_r into the display, $p(x_r)$ and $p'(x_r)$ will be calculated, and the program will stop with $p(x_r)$ in the display. However, x_r and $p'(x_r)$ are stored at various places. So it is very easy to calculate x_{r+1} by (5. 3) and put it into the display. Program II. 4 ends with a command to go back to the beginning. Take that off and put R/S. That will cause a halt, so that one can take time to see if $p(x_r)$ seems near enough zero. After the R/S, put instructions to calculate x_{r+1} by (5. 3) and then an instruction to go back to the beginning of Problem II. 4. After the stop at the end of Program II. 4, with $p(x_r)$ in the display, press R/S. Then x_{r+1} will be calculated, and appear in the display. Press R/S again, and Program II. 4 will run again, giving $p(x_{r+1})$ in the display. Press R/S again, and x_{r+2} will be calculated and appear in the display. With x_r, $p(x_r)$, x_{r+1}, $p(x_{r+1})$,... appearing consecutively, one has some basis for deciding when to stop.

Program II. 4 was written in case we wish actually to know the values of p(c) and

$p'(c)$. But suppose that all we wish is to get a zero by Newton's method. Then we can deal with polynomials of one higher degree. The idea is that we divide all the coefficients of p by a_0. This will not change the zero. But now we can change Program II. 4 a bit to take account of the fact that the leading coefficient is unity; this frees a space to store one more coefficient. The result is embodied in Program VII. 2 in the Program Appendix.

In the next several problems, get the answer to the maximum precision available on your calculator. These problems can be done with a nonprogrammable calculator, but the later ones are really intended for programmable calculators using the program in the Program Appendix.

Problem 5. 3. By Newton's method, find the zero that is near 0 of the polynomial, p, defined by

$$p(x) = x^4 - 4x^3 + 2x^2 - 4x + 1 . \tag{5.17}$$

Problem 5. 4. By Newton's method, find the zero of the p of (5. 17) that is near 4.

Remark. In the two previous problems, we learned that

$$\rho_1 \cong 0.26794\ 91924$$

$$\rho_2 \cong 3.7320\ 50808$$

are two zeros of p. So, by Cor. 2 to Thm. 2. 1 of Chap. II, $(x - \rho_1)(x - \rho_2)$ must divide $p(x)$. By calculator

$$(x - \rho_1)(x - \rho_2) = x^2 - 4x - 1 .$$

Problem 5. 5. By multiplying it out, verify that

$$x^4 - 4x^3 + 2x^2 - 4x + 1 = (x^2 - 4x + 1)(x^2 + 1) . \tag{5.18}$$

Problem 5. 6. By Newton's method, find a zero of the polynomial, p, defined by

$$p(x) = 3x^5 - 9x^4 + 10x^3 + 3x + 1 . \tag{5.19}$$

Problem 5. 7. By Newton's method, find all four zeros of the polynomial, p, defined by

$$p(x) = 8x^4 - 14x^3 - 9x^2 + 11x - 1 . \tag{5.20}$$

Hint. The graph of this is given in Fig. 3. 2 of Chap. II. From this, one can read off reasonable approximations for the four zeros. Find first the zero near 0. 1. Then deflate to a third degree polynomial by the methods given near the end of Sect. 2 of

Chap. II. For it, find the zero near 0.7. Deflate again to a quadratic. Get the last two zeros by the quadratic formula, which is (3.1) of Chap. IV.

Problem 5.8. By Newton's method, find constants c_1, c_2, a, and b so that

$$x^4 - 3x^3 - 4x - 1 = (x - c_1)(x - c_2)(x^2 + ax + b). \tag{5.21}$$

Hint. Proceed as for Prob. 5.7. Find a zero, c_1, and deflate to a third degree polynomial. Find a second zero, c_2, and deflate to a quadratic, which will be $x^2 + ax + b$.

Remark. By the quadratic formula, the zeros of the $x^2 + ax + b$ that you get turn out to be complex, involving imaginary numbers. So $x^2 + ax + b$ is not zero for any real x. So the polynomial expression on the left of (5.21) has only two real zeros, namely c_1 and c_2.

Problem 5.9. Show that the polynomial, p, defined by

$$p(x) = x^3 - 3x^2 + x - 1 \tag{5.22}$$

has one and only one real zero, and find its value to high accuracy.

Hint. Find a zero, c, of (5.22) to high accuracy by Newton's method. Then deflate, getting a and b so that

$$x^3 - 3x^2 + x - 1 = (x - c)(x^2 + ax + b). \tag{5.23}$$

By the quadratic formula, the zeros of $x^2 + ax + b$ turn out to be complex, so that it is not zero for any real x.

Problem 5.10. Find all real zeros of the polynomial, p, defined by

$$p(x) = x^3 + 12x - 4. \tag{5.24}$$

6. Square roots.

To find $\sqrt{A}$, with $A \geq 0$, one can attempt to solve

$$f(x) = x^2 - A = 0 \tag{6.1}$$

by Newton's method. This gives, by (5.3),

$$x_{r+1} = x_r - \frac{x_r^2 - A}{2x_r},$$

since $f'(x) = 2x$. So we have

(6. 2) $$x_{r+1} = \frac{1}{2}(x_r + \frac{A}{x_r}) .$$

Problem 6. 1. Show that if $x_0 > 0$ and $A \geq 0$, then for $r \geq 1$,

(6. 3) $$x_r \geq \sqrt{A} .$$

Hint. Find the minimum point for positive x of the curve

$$y = \frac{1}{2}(x + \frac{A}{x}) .$$

From (6. 2), we have

$$x_{r+1} - \sqrt{A} = \frac{1}{2}(x_r - 2\sqrt{A} + \frac{A}{x_r}) .$$

So

(6. 4) $$x_{r+1} - \sqrt{A} = \frac{1}{2x_r}(x_r - \sqrt{A})^2 .$$

(Incidentally, this is a special case of (5. 7).) This gives

$$x_{r+1} - \sqrt{A} = \frac{x_r - \sqrt{A}}{2x_r}(x_r - \sqrt{A}) \leq \frac{1}{2}(x_r - \sqrt{A}) .$$

So the error is diminished by at least $1/2$ at each step. If x_r is much larger than $\sqrt{A}$, this is what will happen. However, when x_r gets close to $\sqrt{A}$, the error will decrease much faster. If $A \geq 0.25$, then by (6. 3) $x_r \geq 0.5$. Hence by (6. 4)

(6. 5) $$x_{r+1} - \sqrt{A} \leq (x_r - \sqrt{A})^2 .$$

Suppose at x_r we have three decimal places of accuracy. Then

$$x_r - \sqrt{A} \leq 10^{-3} .$$

So by (6. 5)

$$x_{r+1} - \sqrt{A} \leq 10^{-6} .$$

That is, at x_{r+1} we have at least six decimal places of accuracy. So by (6. 5) we can conclude that, for the formula (6. 2) the number of decimal places of accuracy doubles with each successive iteration if $A \geq 0.25$ and $x_0 > 0$.

For large x_r, one has to understand this last remark appropriately. Take $A = 4$ and $x_0 = 10$. Then, by (6.2), $x_1 = 5.2$. However, $\sqrt{4} = 2$, so that both x_0 and x_1 have zero decimal places of accuracy. As twice zero is zero, the number of decimal places of accuracy did double.

Problem 6.2. Use (6.2) to approximate $\sqrt{2}$, and verify that the number of decimal places of accuracy doubles with each successive iteration.

Remark. Formula (6.2) is the one used internally to calculate $\sqrt{A}$ in most calculators because it converges so fast. Different calculators have varied ways to choose x_0. A common way is to choose B, with $0.1 \le B \le 10$, so that $A = B \times 100^r$, for r an integer. Then $\sqrt{B}$ can be determined with five or six applications of (6.2), taking $x_0 = B$. Then $\sqrt{A}$ is taken to be $\sqrt{B} \times 10^r$.

Problem 6.3. Set up Newton's method to solve

$$x^3 - A = 0 ,$$

and so derive a formula similar to (6.2) for approximating $\sqrt[3]{A}$.

7. Integration by partial fractions.

It is explained in the calculus text that if f and g are polynomials and one wishes to evaluate

$$\int \frac{f(x)}{g(x)}\, dx \tag{7.1}$$

by the method of partial fractions, one has to know the factors of $g(x)$. This involves knowing the zeros of g, since by Cor. 1 for Thm. 2.1 of Chap. II, $x - c$ is a factor of $g(x)$ if and only if c is a zero of g.

The exact procedure is explained in detail in the calculus text. See Sect. 7-6 of T-F, for example. So we will not explain it here. But let us give an illustration.

Say we wish to evaluate

$$\int R(x)\, dx , \tag{7.2}$$

where

$$R(x) = \frac{x^4}{x^4 - 4x^3 + 2x^2 - 4x + 1} . \tag{7.3}$$

First, we have to get the numerator to be of lower degree than the denominator. But evidently we have

7. Integration by partial fractions

$$(7.4) \qquad R(x) = 1 + \frac{4x^3 - 2x^2 + 4x - 1}{x^4 - 4x^3 + 2x^2 - 4x + 1} .$$

Now by (5.18), we have

$$\frac{4x^3 - 2x^2 + 4x - 1}{x^4 - 4x^3 + 2x^2 - 4x + 1} = \frac{4x^3 - 2x^2 + 4x - 1}{(x^2 - 4x + 1)(x^2 + 1)} .$$

So we separate into partial fractions

$$(7.5) \qquad \frac{4x^3 - 2x^2 + 4x - 1}{x^4 - 4x^3 + 2x^2 - 4x + 1} = \frac{A}{x - 2 + \sqrt{3}} + \frac{B}{x - 2 - \sqrt{3}} + \frac{Cx + D}{x^2 + 1} .$$

By (7.4) and the theory in the calculus text, the value of (7.2) is

$$(7.6) \qquad x + A \ln(x - 2 + \sqrt{3}) + B \ln(x - 2 - \sqrt{3})$$

$$+ \frac{C}{2} \ln(x^2 + 1) + D \tan^{-1} x + K ,$$

where K is the constant of integration. So all we have to do is find the values of A, B, C, and D.

Multiplying out (7.5) gives

$$(7.7) \qquad 4x^3 - 2x^2 + 4x - 1 = A(x - 2 - \sqrt{3})(x^2 + 1)$$

$$+ B(x - 2 + \sqrt{3})(x^2 + 1)$$

$$+ (Cx + D)(x - 2 - \sqrt{3})(x - 2 + \sqrt{3}) .$$

If we take

$$x = 2 - \sqrt{3} \cong 0.26794\ 91924$$

then the terms on the right of (7.7) that involve B, C, and D all are zero, and we get

$$0.00515\ 47760 \cong A(-2\sqrt{3})(1.0717\ 96770)$$

(we got the value on the left side of (7.7) by Horner's method, of course). So

(7.8) $$A \cong -0.00138\ 83748\ .$$

Now take

$$x = 2 + \sqrt{3} \cong 3.7320\ 50808\ .$$

Then the terms on the right of (7.7) that involve A, C, and D all are zero, and we get

$$193.99\ 48453 \cong B(2\sqrt{3})(14.928\ 20323)\ .$$

So

(7.9) $$B \cong 3.7513\ 88375\ .$$

Then we can write (7.7) as

(7.10) $$4x^3 - 2x^2 + 4x - 1 \cong$$

$$-0.00138\ 83748(x - 3.7320\ 50808)(x^2 + 1)$$

$$+ 3.7513\ 88375(x - 0.26794\ 91924)(x^2 + 1)$$

$$+ (Cx + D)(x^2 - 4x + 1)\ .$$

Put $x = 0$, and get

$$-1 \cong (0.00138\ 83748)(3.7320\ 50808)$$

$$-(3.7513\ 88375)(0.26794\ 91924) + D\ .$$

So

(7.11) $$D \cong 0\ .$$

Finally put $D = 0$ and $x = 1$ in (7.10), and get

(7.12) $$C \cong 0.25\ .$$

Actually, since everything in (7.7) involves small integers or simple multiples of $\sqrt{3}$, it is not too hard to work out the exact values. They are

$$A = \frac{45 - 26\sqrt{3}}{24}$$

$$B = \frac{45 + 26\sqrt{3}}{24}$$

$$C = \frac{1}{4}$$

$$D = 0 .$$

However, make the problem just a shade more difficult and it would be practically impossible without a calculator.

Problem 7. 1. Evaluate

(7. 13)
$$\int \frac{(x^2 - x - 1)\,dx}{8x^4 - 14x^3 - 9x^2 + 11x - 1} .$$

Hint. Look up your answers for Prob. 5. 7.

Problem 7. 2. Evaluate

(7. 14)
$$\int \frac{dx}{x^4 - 3x^3 - 4x - 1} .$$

Hint. Look up your answer for Prob. 5. 8.

8. Linear differential equations with constant coefficients.

The typical linear differential equation with constant coefficients (see Sect. 18-11 of T- F) has the form

(8. 1)
$$a_0 \frac{d^n y}{dx^n} + a_1 \frac{d^{n-1} y}{dx^{n-1}} + \ldots + a_{n-1} \frac{dy}{dx} + a_n y = f(x) .$$

To solve this, one finds somehow ONE solution of (8. 1) and then adds the general solution of

(8. 2)
$$a_0 \frac{d^n y}{dx^n} + a_1 \frac{d^{n-1} y}{dx^{n-1}} + \ldots + a_{n-1} \frac{dy}{dx} + a_n y = 0 .$$

To get the general solution of (8. 2), one tries

(8. 3)
$$y = e^{\rho x} .$$

Substituting this into (8. 2) gives

$$a_0\rho^n e^{\rho x} + a_1\rho^{n-1}e^{\rho x} + \ldots + a_{n-1}\rho e^{\rho x} + a_n e^{\rho x} = 0$$

to be solved. This reduces to

$$\{a_0\rho^n + a_1\rho^{n-1} + \ldots + a_{n-1}\rho + a_n\}e^{\rho x} = 0.$$

If we can find a value of ρ that makes the expression in the curly brackets equal to zero, then, for that value of ρ, (8. 3) gives a solution of (8. 2).

In other words, we are reduced to finding zeros of p, where p is the polynomial defined by

(8. 4) $$p(x) = a_0x^n + a_1x^{n-1} + \ldots + a_{n-1}x + a_n .$$

This is called the characteristic equation for (8. 2).

Given several ρ's that are zeros of p, we can put them into (8. 3) and get several solutions of (8. 2). Where to go from there is explained in the calculus text, to which we refer the reader. But let us look at an illustration.

Say we wish to solve

(8. 5) $$\frac{d^4y}{dx^4} - 4\frac{d^3y}{dx^3} + 2\frac{d^2y}{dx^2} - 4\frac{dy}{dx} + y = 0 .$$

Recall (5. 18), which gave

$$x^4 - 4x^3 + 2x^2 - 4x + 1 = (x - 2 + \sqrt{3})(x - 2 - \sqrt{3})(x^2 + 1).$$

So evidently we can take $\rho_1 = 2 + \sqrt{3}$ and $\rho_2 = 2 - \sqrt{3}$ in (8. 3) to get solutions of (8. 5). What about the factor $x^2 + 1$? If one is not allergic to imaginary numbers, one can use the quadratic formula (which is (3. 1) of Chap. IV) to get two zeros of it, namely $x = \pm i$. So try $\rho = i$ and $\rho = -i$ in (8. 3). If one has two solutions of (8. 5), then a constant times their sum is also a solution. So try

$$y = \frac{1}{2}(e^{ix} + e^{-ix}) .$$

It happens that this equals $\cos x$. You will have no trouble verifying that $y = \cos x$ is a solution of (8. 5). Or try

$$y = \frac{1}{2i}(e^{ix} - e^{-ix}).$$

It happens that this equals $\sin x$. You will have no trouble verifying that $y = \sin x$ is a solution of (8. 5).

Problem 8. 1. Find four solutions of the form (8. 3) for

$$8\frac{d^4y}{dx^4} - 14\frac{d^3y}{dx^3} - 9\frac{d^2y}{dx^2} + 11\frac{dy}{dx} - y = 0 . \tag{8. 6}$$

Hint. Recall your solution of Prob. 5. 7.

Problem 8. 2. Find four solutions of the form (8. 3) for

$$\frac{d^4y}{dx^4} - 3\frac{d^3y}{dx^3} - 4\frac{dy}{dx} - y = 0 . \tag{8. 7}$$

Hint. Recall your solution of Prob. 5. 8. For the quadratic factor, get two zeros involving imaginary numbers by using the quadratic formula. As explained in the calculus text, these two zeros can be used to introduce some cosine and sine terms into the solution.

Chapter VIII

LIMITS AND CONVERGENCE

0. Guide for the reader.

Calculus depends heavily on the theory of limits. Derivatives are defined as limits. Integrals are defined as limits. One is frequently called upon to find the limit of some expression.

For this reason, most calculus texts take time out after a bit to explain the main characteristics of limits. For example, see Sect. 1-10 of T-F. After this explanation in your text, it is advisable to read Sect. 1.

You will find that various calculations in this section will help you understand what happens when a function approaches a limit. The calculator makes it easy for you to carry out numerical experiments, to see what the limiting value might be. On the other hand, the values given by calculators are mostly a trifle in error. Sometimes they are more than a trifle in error, as when cancellation occurs, which is frequent in the kind of limit problems encountered in differentiation. So the calculator will help you understand limits, but for getting the exact value of a limit you must rely on calculus.

Often you do not need the exact value. Then the calculator can often be very helpful. One instance of this is a way of getting a root of an equation by a limiting process called fixed point iteration. This is discussed in Sect. 2. This process is quite simple, and you may often find it more attractive, or easier to program, than the methods given in Chap. VII for finding a root of an equation.

In Sect. 3, we discuss the Aitken δ^2 process as a simple example of a method for accelerating the convergence of a sequence. This section gives you an example of how to exploit convergence behavior of a sequence, as established by calculus, to get to the limit more quickly.

1. Numerical illustrations.

Can the calculator be of help in finding limits? Let us look at a few examples. In order to find out what the derivative of $\sin x$ is, one has to show that

1. Numerical illustrations

$$\lim_{\theta \to 0} \frac{\sin \theta}{\theta} = 1 \tag{1.1}$$

(PROVIDED that θ is taken in radians). This result is (1) in Sect. 2-10 of T-F. In Table 1.1 we give values of $(\sin \theta)/\theta$ for values of θ tending to zero. These were calculated on the HP-33E, but other calculators give similar results. This looks pretty

Table 1.1

θ	$\frac{\sin \theta}{\theta}$
± 1	0.84147 09848
$\pm 10^{-2}$	0.99998 33334
$\pm 10^{-4}$	0.99999 99983
$\pm 10^{-6}$	1.00000 00000
$\pm 10^{-8}$	1.00000 00000
$\pm 10^{-10}$	1.00000 00000

convincing, provided you feel that the HP-33E (or your own calculator) is doing an accurate job of calculating $\sin \theta$.

However, it does not PROVE anything. For all we know, if we could take $\theta = 10^{-433}$, which is beyond the range of the calculator, $(\sin \theta)/\theta$ might come out to be 0.6. Or it could be that the limit really is $1 + 10^{-27}$, which would accord perfectly with the 10-digit values in Table 1.1.

Actually, your calculator is probably rigged to perform this limit calculation perfectly. On some calculators, $\sin \theta$ is calculated for θ near 0 from a formula of the form

$$\sin \theta = \theta(1 - \theta^2 g(\theta)),$$

with g a well behaved function (very likely a rational function) for which

$$\lim_{\theta \to 0} g(\theta) = \frac{1}{6}.$$

Thus for θ close enough to 0, such a calculator returns θ itself as the value for $\sin \theta$. For example, on a 10-digit calculator, like the HP-33E, this is bound to happen when $|\theta| \leq 1.7 \times 10^{-5}$. Then we have

$$|\theta^2 g(\theta)| \cong |\theta^2 g(0)| < \frac{10^{-10}}{2},$$

so that subtraction of $\theta^2 g(\theta)$ from 1 leaves 1 unchanged. On other calculators, $\sin\theta$ is obtained as $(\tan\theta)/\sqrt{1+(\tan\theta)^2}$, with $\tan\theta$ calculated in a process which accurately builds up $\tan\theta$ from the accurate tangents of certain selected small angles. This process, too, insures that, for θ near 0, the calculator returns θ (correctly) as the value of $\tan\theta$ and therefore of $\sin\theta$.

But, as soon as the calculator obtains θ as the value for $\sin\theta$, then division by θ is bound to produce the answer 1.

For a related example, consider

$$\lim_{x\to 0} \frac{\sin 5x}{\sin 3x} \quad (1.2)$$

This is Prob. 7 at the end of Sect. 2-10 of T-F.

We have calculated Table 1.2 on the HP-33E. We are sure the reader will recognize the later entries as a very good numerical approximation to 5/3. So the limit is

Table 1.2

x	$\frac{\sin 5x}{\sin 3x}$
± 1	-6.7950 97928
$\pm 10^{-2}$	1.6662 22231
$\pm 10^{-4}$	1.6666 66622
$\pm 10^{-6}$	1.6666 66667
$\pm 10^{-8}$	1.6666 66667
$\pm 10^{-10}$	1.6666 66667

doubtless 5/3, though we certainly have not proved it. By using calculus, one can prove that the limit is indeed 5/3 on the nose.

Problem 1.1. Explain why your calculator is apt to give the ratio M/N to full accuracy when asked to calculate the ratio $(\sin Mx)/(\sin Nx)$ for x very near 0.

Problem 1.2. Find

$$\lim_{x\to 0} \frac{\sin 2x}{2x^2+x} .$$

Remark. This is Prob. 19 at the end of Sect. 2-10 of T-F.

We now consider examples where the calculator is less helpful, such as

$$\lim_{x \to -1} \frac{x^3 + 1}{2x^2 - 3x - 5} . \tag{1.3}$$

Table 1.3 shows some sample calculations on the HP-33E. The "Error" entries arise because for those values of x the denominator comes out to be zero. Calculators will not divide by zero.

Table 1.3

Δx	$\frac{x^3+1}{2x^2-3x-5}$ for $x = -1+\Delta x$
10^{-2}	-0.42551 57593
10^{-4}	-0.42854 08155
10^{-6}	-0.42857 14286
10^{-8}	-0.42857 14286
10^{-9}	-0.42857 14286
10^{-10}	Error
-10^{-10}	Error
-10^{-9}	-0.42857 14286
-10^{-8}	-0.42857 14286
-10^{-6}	-0.42857 14286
-10^{-4}	-0.42860 20399
-10^{-2}	-0.43163 81766

One is not inclined to doubt that the limit is approximately -0.42857 14286. If one is astute enough to recognize that this is a very close approximation to $-3/7$, one will suspect that $-3/7$ is indeed the limit. Numerical answers in calculus are quite often simple fractions!

However, it turns out that the values we chose for Δx were quite special, designed to work unusually well on a decimal digit calculator. To see this, we try some "funny" values of x close to -1. After all, if the limit for (1.3) is really $-3/7$, one should get close to that for any value of x close to -1. But look at the numbers in Table 1.4. The value for $\Delta x = 5 \times 10^{-10}$ casts grave doubt whether the limit could possibly be $-3/7$. On the other hand, look at the values for negative Δx.

What is happening is that we are encountering cancellation, and the value as

Table 1. 4

Δx	$\frac{x^3+1}{2x^2-3x-5}$ for $x = -1 + \Delta x$	
	calculated value	accurate value
1.23459×10^{-5}	-0. 42856 71307	-4. 2856 76492
1.2349×10^{-6}	-0. 42858 63026	-4. 2857 10505
1.239×10^{-7}	-0. 42871 97232	-4. 2857 13906
1.29×10^{-8}	-0. 43000 00000	-4. 2857 14246
1.9×10^{-9}	-0. 43846 15385	-4. 2857 14280
5×10^{-10}	-0. 50000 00000	-4. 2857 14285
-5×10^{-10}	-0. 42857 14286	-4. 2857 14289
-1.9×10^{-9}	-0. 42857 14286	-4. 2857 14292
-1.29×10^{-8}	-0. 42857 14286	-4. 2857 14326
-1.239×10^{-7}	-0. 42857 14286	-4. 2857 14665
-1.2349×10^{-6}	-0. 42857 14286	-4. 2857 18066
-1.23459×10^{-5}	-0. 42857 14286	-4. 2857 52079

given by the calculator is differing considerably from the true value for the given value of x The true value (to 10 digits) is given in the right hand column of Table 1. 4.

Other calculators will exhibit similar phenomena, but not necessarily for the same values of Δx.

Cancellation is discussed in Sect. 2 of Chap. IV. Some possible ways of avoiding it are discussed in Sect. 3 of Chap. IV. In the present case, we can proceed as follows. By Thm. 2. 1 of Chap. II, we have

$$x^3 + 1 = (x^2 - x + 1)(x + 1)$$

$$2x^2 - 3x - 5 = (2x - 5)(x + 1) .$$

So

$$\frac{x^3 + 1}{2x^2 - 3x - 5} = \frac{x^2 - x + 1}{2x - 5} . \tag{1. 4}$$

On the right side of (1. 4), there is no cancellation. This equation was used to calculate the right hand column of Table 4. 1. Indeed, one can put $x = -1$ in the right side of (1. 4) and determine that the limit in (1. 3) is exactly $-3/7$.

Table 1. 5

Δx	$\frac{x^3 + 1}{2x^2 - 3x - 5}$ for $x = -1 + \Delta x$ true value (rounded to 10 digits)	Error in value given in Table 1. 3
10^{-2}	-0. 42551 57593	0
10^{-4}	-0. 42854 08169	-14×10^{-10}
10^{-6}	-0. 42857 11224	3062×10^{-10}
10^{-8}	-0. 42857 14255	31×10^{-10}
10^{-10}	-0. 42857 14286	Error
-10^{-10}	-0. 42857 14286	Error
-10^{-8}	-0. 42857 14316	-30×10^{-10}
-10^{-6}	-0. 42857 17347	-3061×10^{-10}
-10^{-4}	-0. 42860 20414	-15×10^{-10}
-10^{-2}	-0. 43163 81766	0

As we said, the Δx's in Table 1. 3 were quite special. They strongly supported the limit $-3/7$, but this was because the calculator was giving the wrong values. In Table 1. 5, we give the correct values (calculated from the right side of (1. 4), naturally). The disparity between the values in Tables 1. 3 and 1. 5 is quite appreciable.

<u>Moral</u>. Don't believe everything you see on a calculator.

Incidentally, Table 1. 5 gives a much better image of what happens when one has a limit than Table 1. 3 does. In Table 1. 5, one sees that as $x \to -1$ the given expression does indeed approach $-3/7$, but at a steady rate; not abruptly, as one would think from Table 1. 3.

Incidentally, you can also find that the limit in (1. 3) is $-3/7$ by using l'Hôpital's rule. If you differentiate both the numerator and denominator, you get the ratio of derivatives

$$\frac{3x^2}{4x - 3}.$$

Putting $x = -1$ gives $-3/7$.

Problem 1.3. Guess the limit as $x \to 2$ of

$$\frac{x^2 - 4}{x^3 - 8} \tag{1.5}$$

by calculating its value for $x = 2 + \Delta x$ for $\Delta x = \pm 10^{-n}$, $n = 2, 3, \ldots, 9$, on your calculator. Calculate the correct values of (1.5) for the same values of x, and determine the exact limit.

Remark. This is nearly the same as Example 6 in Sect. 1-10 of T-F.

A limit that occurs frequently in calculus is the derivative, defined by

$$f'(x) = \lim_{h \to 0} \frac{f(x+h) - f(x)}{h}. \tag{1.6}$$

In trying to get a numerical approximation for $f'(x)$ by calculator, cancellation is bound to give difficulty. This point is discussed somewhat in Chap. III, and in detail in Chap. V.

Since the result of cancellation depends on the roundoff error, which varies from one calculator to another, different calculators will give quite distinct results if tried on the right side of (1.6) for "small" h. A striking example of this is given in Table 2.2 of Chap. III, where the right side of (1.6) is calculated for $f(x) = x^{-1}$ at $x = 0.7$ on three different calculators. Another example is to be found in Tables 1.1 and 1.2 of Chap. V, where minor differences in h are greatly exaggerated due to cancellation.

For exercises involving this point, see Problems 1.6, 1.7, and 1.8 of Chap. III. It would be instructive in each of these to use still smaller values of Δx than suggested for the problems, since the ranges of Δx for those problems were restricted to avoid difficulty for the student.

Cancellation can arise in many other situations. In Prob. 18 at the end of Sect. 3-9 of T-F, we are asked for the limit as $x \to \infty$ of

$$x - \sqrt{x^2 + x}. \tag{1.7}$$

Some calculator results, as done on the HP-33E and the TI-30, are presented in Table 1.6. To get the true value, without cancellation, we multiply (1.7) by

$$\frac{x + \sqrt{x^2 + x}}{x + \sqrt{x^2 + x}},$$

a trick given in Sect. 3 of Chap. IV. The result is

Table 1. 6

x	$x - \sqrt{x^2 + x}$		
	calculated value		accurate value
	HP-33E	TI-30	
1	-0. 41421 35620	- . 41421 356	-0. 41421 35624
10	-0. 48808 84800	- . 48808 848	-0. 48808 84817
10^2	-0. 49875 62000	-0. 49875 62	-0. 49875 62112
10^3	-0. 49987 50000	-0. 49987 5	-0. 49987 50625
10^4	-0. 49999 00000	-0. 49998	-0. 49998 75006
10^5	-0. 50000 00000	-0. 4999	-0. 49999 87500
10^6	-0. 50000 00000	-0. 499	-0. 49999 98750
10^7	-0. 50000 00000	-0. 49	-0. 49999 99875
10^8	-0. 50000 00000	-0. 4	-0. 49999 99988
10^9	0. 00000 00000	0	-0. 50000 00000
10^{10}	0. 00000 00000	0	-0. 50000 00000

(1. 8)
$$x - \sqrt{x^2 + x} = \frac{-x}{x + \sqrt{x^2 + x}} .$$

If we let $x \to \infty$ on the right side, we get $-1/2$ as the limit.

Observe again that the right column of Table 1. 6, where the true values are given, gives a much more understandable picture of how (1. 7) approaches the limit $-1/2$ than either column of calculated values does. Note also how the last decent value provided by the TI-30 is spoiled by the fact that the TI-30 truncates rather than rounds.

Let us try to find the limit as $\theta \to 0$ of

(1. 9)
$$\frac{1 - \cos \theta}{\theta^2} .$$

Calculated values, as done on the HP-33E and the TI-30, are given in Table 1. 7. For $|\theta| \le 10^{-5}$, the calculator gives $\cos \theta = 1$ to 10 decimals, and so gives 0 for (1. 9). The true values are derived from the trigonometric identity

(1. 10)
$$\frac{1 - \cos \theta}{\theta^2} = \frac{2 \sin^2 (\theta/2)}{\theta^2} .$$

Table 1. 7

θ	$\frac{1-\cos\theta}{\theta^2}$ calculated value: HP-33E	calculated value: TI-30	accurate value
± 1	0. 45969 76941	. 45969 769	0. 45969 76942
$\pm 10^{-1}$	0. 49958 34700	. 49958 344	0. 49958 34722
$\pm 10^{-2}$	0. 49999 60000	. 49999 5	0. 49999 58334
$\pm 10^{-3}$	0. 50000 00000	. 499	0. 49999 99584
$\pm 10^{-4}$	0. 50000 00000	0. 4	0. 49999 99996
$\pm 10^{-5}$	0. 00000 00000	0	0. 50000 00000
$\pm 10^{-6}$	0. 00000 00000	0	0. 50000 00000

From this one can easily conclude that the limit of (1. 9) is exactly $1/2$. One can also establish this result by two successive applications of l'Hôpital's rule.

Problem 1. 4. Find

$$\lim_{y \to \infty} 2y \tan \frac{\pi}{y} . \tag{1. 11}$$

Remark. This is Prob. 15 at the end of Sect. 2-10 of T- F.

Note that you certainly cannot get the exact value of this limit except by calculus, and you will do well even to guess it without calculus.

In summary: When you are trying to find

$$\lim_{x \to x_0} f(x) ,$$

the calculator is of help to you since it makes it easy for you to carry out numerical experiments. You can look at values of $f(x)$ for x closer and closer to x_0 and, in this way, form an impression of what the limiting value might be. It is even possible to be more systematic about these experiments and use extrapolation techniques, such as discussed in Sect. 3. But, numerical experiments cannot establish the existence of the limit nor its exact value. For that, you need calculus.

*2. Roots by fixed point iteration.

In Chap. VII, we discussed various ways to get a root of an equation. We discuss here an attractive method which depends on a simple iteration.

We illustrate the ideas by considering the equation

$$x^4 - 4x^3 + 2x^2 - 4x + 1 = 0 . \tag{2.1}$$

Rewrite this successively as

$$x^4 - 4x^3 + 2x^2 - 4x = -1 ,$$

$$x(x^3 - 4x^2 + 2x - 4) = -1 ,$$

$$x = \frac{-1}{x^3 - 4x^2 + 2x - 4} . \tag{2.2}$$

It was shown in Chap. VII that (2.1) has a root near 0. So let us take

$$x_0 = 0 \tag{2.3}$$

as a first guess for a root. Substitute this into the right side of (2.2). This gives a second guess

$$x_1 = 0.25 . \tag{2.4}$$

If this happens to be a better guess than x_0 was, then substituting it into the right side of (2.2) should give a still better guess.

If we are going to keep substituting things into the right side of (2.2), we had better fix it up to use Horner's method. So we write

$$x_{r+1} = \frac{-1}{((x_r - 4)x_r + 2)x_r - 4} . \tag{2.5}$$

Here we have guessed x_r (for example x_0), and we substitute it into the right side of (2.2), namely the right side of (2.5), to get a new guess x_{r+1} (for example the x_1 that we got before). It is very easy to write a program which starts with x_r in the display, calculates the right side of (2.5), and winds up with x_{r+1} in the display, ready to run the program again.

Let us just try this out for a few r's, as shown in Table 2.1. It looks very much as if the x_r's are approaching a limit, x. We express this by saying that the x_r's are <u>converging</u> to a limit x. If one would continue with (2.5) until that limit x were

Table 2. 1

r	x_r
0	0. 00000 00000
1	0. 25000 00000
2	0. 26778 24268
3	0. 26794 83264
4	0. 26794 91879

reached, one would have x_{r+1} and x_r the same, namely both x, and one could multiply back to get (2. 1). So the x_r's are converging to a root of (2. 1).

Table 2. 1 provides strong evidence (but no proof) that to five decimals a root of (2. 1) is $x \cong 0.26795$.

Problem 2. 1. Set up a sequence of x_r's that converge to the root of (2. 1) that is near 4, and so get a 5 decimal approximation.

Hint. It is no use trying to do anything with (2. 5). This will always give a sequence of x_r's converging to about 0. 26795. (See Prob. 2. 7.) However, take $c = 4$ in Thm. 2. 1 of Chap. II. This will give

$$x^4 - 4x^3 + 2x^2 - 4x + 1 = (x - 4)(x^3 + 2x + 4) + 17 .$$

So we can rewrite (2. 1) successively as

$$(x - 4)(x^3 + 2x + 4) = -17 ,$$

$$x - 4 = \frac{-17}{x^3 + 2x + 4} ,$$

$$x = 4 - \frac{17}{((x + 0)x + 2)x + 4} . \qquad (2.6)$$

So, in place of (2. 5), take

$$x_{r+1} = 4 - \frac{17}{((x_r + 0)x_r + 2)x_r + 4} . \qquad (2.7)$$

Start with $x_0 = 4$. Save your successive x_r's for future use in Sect. 3.

If you have already read Chap. VII, you will recall that Newton's method generates a sequence of x_r's. We did not point it out at that time, but in fact they converge to a zero of a function $f(x)$, that is to a root of the equation

$$f(x) = 0 .$$

In Newton's method, each x_{r+1} is defined in terms of x_r. We have the same sort of thing here; see (2. 5) and (2. 7). In this sort of circumstance, we say that the x_r's are defined by iteration.

Notice that both (2. 2) and (2. 6) have the form

$$x = g(x) . \tag{2. 8}$$

In trying to solve (2. 8), we are seeking an x that is carried into itself when operated upon by the function g. That is, x is a fixed point of g. So we say that we have given a fixed point iteration to find a root of (2. 1).

Problem 2. 2. Determine

$$s = \sqrt[4]{5 + \sqrt[4]{5 + \sqrt[4]{5 + \sqrt[4]{5 + \ldots}}}} \ . \tag{2. 9}$$

Hint. Form $\sqrt[4]{5 + s}$. Putting an extra $\sqrt[4]{5 +}$ on the right side of (2. 9) will not change it any. So we must have

$$\sqrt[4]{5 + s} = s . \tag{2. 10}$$

That is, we wish a fixed point of the g defined by

$$g(x) = \sqrt[4]{5 + x} . \tag{2. 11}$$

So, as we did earlier, let us define a sequence of x_r's by the iteration

$$x_{r+1} = \sqrt[4]{5 + x_r} . \tag{2. 12}$$

Problem 2. 3. Determine the limit of the sequence of x_r's determined by the fixed point iteration

$$x_{r+1} = \sqrt{x_r} , \tag{2. 13}$$

with x_0 a positive number.

Problem 2. 4. Vieta's formula

$$\frac{2}{\pi} = \sqrt{\frac{1}{2}} \times \sqrt{\left(\frac{1}{2}+\frac{1}{2}\sqrt{\frac{1}{2}}\right)} \times \sqrt{\left(\frac{1}{2}+\frac{1}{2}\sqrt{\left(\frac{1}{2}+\frac{1}{2}\sqrt{\frac{1}{2}}\right)}\right)} \times \ldots$$

(see Sect. 16-10 of T-F), is apparently of the form

$$\frac{2}{\pi} = a_1 a_2 a_3 \cdots$$

with the a_i's obtained by fixed point iteration,

$$a_{n+1} = \sqrt{\frac{1+a_n}{2}}, \quad n = 0, 1, 2, \ldots, \quad a_0 = 0 .$$

Show that $\lim_{n \to \infty} a_n = 1$ and see how accurately this formula gives

$$\frac{2}{\pi} = 0.63661\ 97723\ 67581\ 34307\ldots$$

on your calculator.

Remark. You may stop the iteration as soon as your calculator obtains an a_n which is exactly 1 (after which the computed product $p_n = a_1 a_2 \ldots a_n$ will not change any more with n). This will happen fairly fast. For example, $a_{12} = 1$ on an eight digit calculator, while $a_{16} = 1$ on a ten digit calculator.

It was shown in Chap. VII that the equation

$$x = \ln x + 2 \tag{2.14}$$

has two roots. They would be fixed points of the function g defined by

$$g(x) = \ln x + 2 .$$

Problem 2.5. Set up the obvious fixed point iteration to generate a sequence of x_r's that converge to a root of (2.14), and so get what appears to be an approximation to five significant digits. If you have a nonprogrammable calculator, save your last four x_r's for future use in Sect. 3.

Problem 2.6. Find to five significant digits the other root of (2.14) by a fixed point iteration. If you have a nonprogrammable calculator, save your last four x_r's for future use in Sect. 3.

Hint. As you can easily find out by experiment, the obvious fixed point iteration for (2.14) leads always to the same root, no matter where you start. Rewrite (2.14) as

$$\ln x = x - 2 .$$

Take the exponential of each side, and get

$$x = e^{x-2} . \tag{2.15}$$

Now try the obvious fixed point iteration for (2.15).

Incidentally, it is easy to understand fixed point iteration with just a bit of calculus. (See Prob. 4 of the miscellaneous problems for Chap. 16 of T-F.) Given that c is a fixed point of the iteration function g, i.e.,

$$g(c) = c \ ,$$

while $x_{r+1} = g(x_r)$, you get for the errors that

$$c - x_{r+1} = g(c) - g(x_r) \ .$$

But, assuming that g is differentiable, the Mean Value Theorem tells you that

$$g(c) - g(x_r) = g'(\xi)(c - x_r)$$

for some ξ between c and x_r. Consequently, if $|g'(x)| < 1$ for all x near c, and, in particular, for all x between c and x_r, then x_{r+1} is closer to c than is x_r.

As a special case, assume that $|g'(c)| < 1$ and that g' is continuous. Then $|g'(x)| < 1$ for all x close to c. But then, once you have an x_r close enough to c, x_{r+1} will be even closer, and so x_{r+2} will be even closer than that, etc. In short, under those circumstances, we get convergence to c provided we start close enough to c.

By the same token, if $|g'(c)| > 1$, then for all x "near" c, we have $|g'(x)| > 1$, and then, even if x_r is very close to c, x_{r+1} is bound to be <u>farther</u> away. There is then no chance to get close to such a fixed point c by iteration.

This is exactly what happens in Prob. 2.6, if you try to find the fixed point near 0.1 of the function g given by

$$g(x) = \ln x + 2 \ .$$

As you probably know by now, $g'(x) = 1/x$, and this is greater than 1 for all positive x less than 1. So there is no hope of having fixed point iteration converge to that fixed point with g as the iteration function.

<u>Problem 2.7.</u> Show that iteration with (2.5) has no chance of converging to the root of (2.1) near 4.

*3. Aitken's δ^2 process

The common way to try to find a fixed point of g, namely an x such that

$$x = g(x) \ , \tag{3.1}$$

is to set up a fixed point iteration

(3. 2) $$x_{r+1} = g(x_r), \qquad r = 0, 1, 2, \ldots,$$

and hope that it converges. Is there any way to speed up the convergence? There often is, and here is a way to do it.

Observe that trying to solve (3. 1) is the same as trying to find a point (x, y) at which the curves

(3. 3) $$y = x$$

(3. 4) $$y = g(x)$$

cross each other. If we have three consecutive x_r's, say x_r, x_{r+1}, and x_{r+2}, we can estimate the x value of the point where the curves cross. This is done as follows. By (3. 2), the points (x_r, x_{r+1}) and (x_{r+1}, x_{r+2}) both lie on the curve $y = g(x)$. See Fig. 3. 1. So we can find the equation of the secant through these two points. Then we

Figure 3. 1

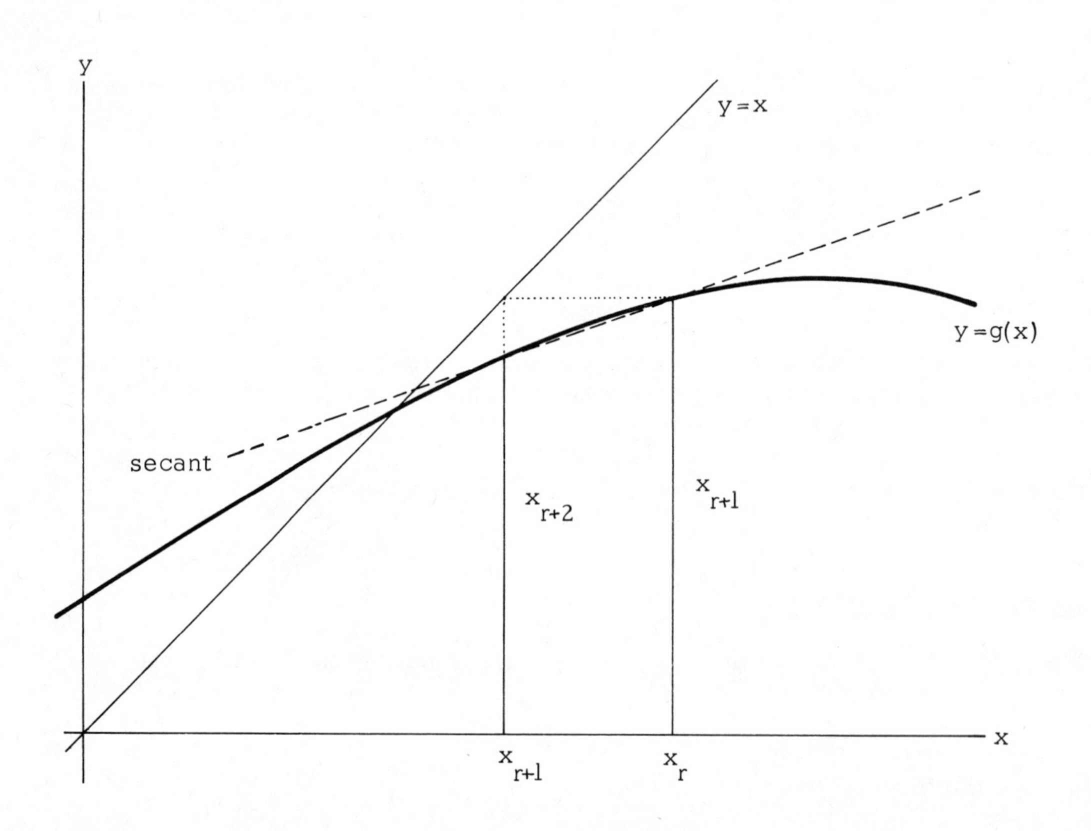

can find the x^* where the secant meets the straight line $y = x$. But (look again at Fig. 3.1), we hope that is near the x where $y = x$ and $y = g(x)$ cross.

The algebra for this works out as follows. The slope of the secant is

$$\frac{x_{r+2} - x_{r+1}}{x_{r+1} - x_r} . \tag{3.5}$$

The equation of the secant is

$$y - x_{r+2} = \frac{x_{r+2} - x_{r+1}}{x_{r+1} - x_r}(x - x_{r+1}) . \tag{3.6}$$

To find the x^* where this crosses $y = x$, we put $y = x$ in (3.6), and solve for x^*. This gives

$$x^* = \frac{x_{r+2}x_r - (x_{r+1})^2}{x_{r+2} - 2x_{r+1} + x_r} . \tag{3.7}$$

An attempt to calculate by this formula will certainly result in serious cancellation, expecially when x_r, x_{r+1}, and x_{r+2} are close to the limit. But we can write (3.7) in other forms, such as

$$x^* = x_{r+2} - \frac{(x_{r+2} - x_{r+1})^2}{x_{r+2} - 2x_{r+1} + x_r} . \tag{3.8}$$

The last term on the right is still subject to cancellation. However, we hope that x_{r+2} is fairly close to x^*. So the final term is small compared to x_{r+2}. Even if this correction term cannot be accurately calculated, due to cancellation, an inaccurate value will usually suffice to give x^* to high accuracy. Specifically, suppose x_{r+2} is already close enough to the limit that the first five digits are correct. Then only the final five digits need to be corrected. For this, one needs the correction term with an accuracy of only five digits.

Indeed, (3.7) and (3.8) can be put in still another form, even less subject to cancellation error, specifically

$$x^* = x_{r+2} + (x_{r+2} - x_{r+1}) \div \left\{ \frac{x_{r+1} - x_r}{x_{r+2} - x_{r+1}} - 1 \right\} . \tag{3.9}$$

RPN
|
RPN

Here is a program for calculating x^* which uses only one memory register:

RPN

(3.10) x_{r+2}, [↑], x_{r+1}, [↑], [↑], x_r, [−], [STO],

[R↓], [−], [↑], [↑], [RCL], [x⇄y], [÷], 1, [−],

[÷], [+] .

The first line calculates and stores $x_{r+1} - x_r$. The second line calculates the curly bracket in (3.9).

If the calculator has more than one memory register, some of the x_r's can be stored, and some (or all) of the inputs in (3.10) can be replaced by recalls. This latter form can be easily stored as a program, if one has a programmable calculator.

RPN

AE

Here is a program for calculating x^* which uses only one memory register:

(3.10) x_{r+2}, [−], x_{r+1}, [=], [STO],

x_{r+1}, [−], x_r, [=], [÷], [RCL], [−], 1, [=],

[1/x], [×], [RCL], [=], [+], x_{r+2}, [=] .

The first line calculates and stores $x_{r+2} - x_{r+1}$. The second line calculates the curly bracket in (3.9).

If your calculator has an EXC or x⇄t key which exchanges the contents of some memory register with the x- or display register, you should use it here to avoid inputting x_{r+1} twice. Put [STO] right after the first input of x_{r+1}, and replace the two adjacent steps [STO], x_{r+1} by [EXC]. If your calculator has a parenthesis capability, then, even if your calculator has only one memory register, it is possible to avoid inputting either x_{r+1} or x_{r+2} twice. This is done by evaluating (3.9) from left to right, as follows:

x_{r+2}, [+], [(], [(], [STO], [−], x_{r+1}, [STO], [)], [÷],

[(], [(], [CHS], [÷], [(], x_r, [−], [RCL], [)], [)], [1/x], [−], 1, [=] .

The first [STO] in the program is irrelevant to the calculation. At the point where it is actuated, x_{r+2} is in the display, so that the subsequent [−] and x_{r+1} should produce $x_{r+2} - x_{r+1}$. However, this is just after a [(]. On some makes of calculators, this renders the x_{r+2} inactive. However, the [STO] restores it to an active role. Similarly for the change sign instruction, [CHS], that appears later.

If the calculator has more than one memory register, some of the x_r's can be stored, and some (or all) of the inputs in (3.10) can be replaced by recalls. This latter form can be easily stored as a program if one has a programmable calculator.

AE

3. Aitken's δ^2 process

In Table 3. 1, we show the results of applying Program (3. 10) to the results in Table 2. 1. Under x_r in Table 3. 1 we merely repeat the entries of Table 2. 1. Under

Table 3. 1

r	x_r	x^*
0	0. 00000 00000	
1	0. 25000 00000	
2	0. 26778 24268	0. 26914 41442
3	0. 26794 83264	0. 26794 98887
4	0. 26794 91879	0. 26794 91924

x^* we give the values from using (3. 9), as applied to three consecutive entries under x_r. The x^* opposite $r = 2$ is poor, being in fact further from the true root than x_2 itself. This means that we are far enough from the root that the secant (in Fig. 3. 1) does not hit the line $y = x$ particularly close to where $y = g(x)$ hits it. For the next x^*, we are closer, and do better. The final x^* agrees to <u>all</u> 10 decimals with the exact root of (2. 1), namely $2 - \sqrt{3}$. Incidentally, if you wish to check this, do not subtract $\sqrt{3}$ from 2 on your calculator. This gives cancellation in calculating $2 - \sqrt{3}$, and one significant digit is lost. Use instead $1/(2 + \sqrt{3})$, which (mathematically) is the same as $2 - \sqrt{3}$.

Use of (3. 9) to improve the approximation to a limit is called the <u>Aitken's δ^2 process.</u>

<u>Problem 3. 1.</u> Use the Aitken's δ^2 process with the results of Prob. 2. 1 to get an approximation to about 10 digit accuracy for the root of (2. 1) that is near 4.

<u>Remark.</u> The root in question happens to be $2 + \sqrt{3}$.

<u>Problem 3. 2.</u> Use the Aitken's δ^2 process to get approximations to about 10 digit accuracy for the roots of (2. 14). Use the results of Prob. 2. 5 for the root greater than 1 and the results of Prob. 2. 6 for the root less than 1.

<u>Remark.</u> 10 digit approximations for the two roots in question are 3. 1461 93220 and 0. 15859 43396.

The Aitken's δ^2 process is at times effective in accelerating convergence of a sequence of x_r's even if they are not generated by a fixed point iteration. This is the case if the x_r's approach a limit L and there are constants a and k, with $|k| < 1$, such that

$$x_r \cong L + ak^r . \tag{3.11}$$

Since $|k| < 1$, we have

$$\lim_{r \to \infty} x_r = L .$$

We raise the question: Given values of a few x_r's, what can we do about guessing L? If (3.11) were exact instead of approximate, we could solve exactly for L from three consecutive values of x_r. We have

(3.12)
$$x_{r+1} - x_r = ak^r(k - 1) ,$$

(3.13)
$$x_{r+2} - x_{r+1} = ak^{r+1}(k - 1) .$$

Hence

$$x_{r+2} - 2x_{r+1} + x_r = ak^r(k - 1)^2 .$$

So

$$\frac{(x_{r+2} - x_{r+1})^2}{x_{r+2} - 2x_{r+1} + x_r} = ak^{r+2} .$$

If we use $r + 2$ in place of r in (3.11), and substitute from the last equation above, we get

(3.14)
$$L = x_{r+2} - \frac{(x_{r+2} - x_{r+1})^2}{x_{r+2} - 2x_{r+1} + x_r} .$$

This can be rewritten as

(3.15)
$$L = x_{r+1} + (x_{r+2} - x_{r+1}) \div \left\{ \frac{x_{r+1} - x_r}{x_{r+2} - x_{r+1}} - 1 \right\} .$$

Since (3.11) was only approximate, the same holds for (3.14) and (3.15). However, by comparing (3.14) with (3.8), we see that if the behavior of x_r is governed by (3.11), then Aitken's δ^2 process is useful for estimating the limit L.

Consider again our problem of finding the limit as $\theta \to 0$ of

(3.16)
$$\frac{1 - \cos\theta}{\theta^2} .$$

If one takes successively $\theta = 2, 1, 0.5, 0.25, 0.125, 0.0625, \ldots$, each θ being half the one before it, in (3.16) one should get a sequence of values that converges to $1/2$, which is the limit of (3.16). In Table 3.2, we give some results of this, calculating L by (3.15) with the HP-33E.

Table 3. 2

θ	$(1-\cos\theta)/\theta^2$ calculated	L	R
2	0. 35403 67093		
1	0. 45969 76941		
0. 5	0. 48966 97524	0. 50153 83879	3. 525
0. 25	0. 49740 12528	0. 50008 89596	3. 877
0. 125	0. 49934 92992	0. 50000 54616	3. 969
0. 0625	0. 49983 72608	0. 50000 03380	3. 992
0. 03125	0. 49995 92960	0. 49999 99943	3. 999
0. 015625	0. 49998 97088	0. 49999 98039	4. 013
0. 0078125	0. 49999 70816	0. 49999 94409	4. 125

The values of L are considerably better than the corresponding values to their left. Recall that the limiting value is 1/2. Note, though, that the last two extrapolated values are not as good as their precedessors. This illustrates a computational difficulty with extrapolation to the limit. Any errors in the calculated values for x_r, x_{r+1}, and x_{r+2} will usually be magnified by the extrapolation process, because of the cancellation which occurs in the evaluation of (3. 15). These errors don't matter as long as L itself is not too accurate an approximation to the limit. But, in our example, as θ comes close to zero, the terms in our sequence have increasing relative error because of cancellation in their calculation, and these errors in turn spoil the accuracy of the extrapolated value L. This can be documented here by carrying the calculation out again using accurate values for $(1 - \cos\theta)/\theta^2$, as calculated from (1. 10). The results are shown in Table 3. 3.

Table 3. 3

θ	$(1-\cos\theta)/\theta^2$ accurate	L
2	0. 35403 67092	
1	0. 45969 76942	
0. 5	0. 48966 97526	0. 50153 83882
0. 25	0. 49740 12528	0. 50008 89594
0. 125	0. 49934 92972	0. 50000 54580
0. 0625	0. 49983 72608	0. 50000 03398
0. 03125	0. 49995 93108	0. 50000 00204
0. 015625	0. 49998 98274	0. 50000 00014
0. 0078125	0. 49999 74568	0. 50000 00000

There are ways of monitoring the convergence behavior of the extrapolated values which allow you at time to pinpoint the moment when noise begins to spoil the accuracy of the extrapolated value. The key lies in observing the ratios

$$R_r = \frac{x_{r+1} - x_r}{x_{r+2} - x_{r+1}} .$$

which occur explicitly in (3. 9) and (3. 15) and which, under ideal circumstances, converge to $1/k$, as can be seen from (3.11)- (3. 13). If the sequence comes from fixed point iteration, then these ratios have geometric meaning. Indeed, by (3. 5), R_r is the reciprocal of the slope of the secant in Fig. 3. 1. So R_r should approach $1/g'(c)$, where c is the desired solution. In calculations, these ratios will settle down initially, but eventually, as cancellation becomes more severe in the evaluation of (3. 15), the ratios will become noisy. This is shown in Table 3. 2, where we have also listed the ratios R_r as well. Note how the difference between successive ratios first decreases, but eventually becomes larger again. This is a clear indication that noise has become an important part of the computed extrapolated values.

Problem 3. 3. Fine balancing scales in chemistry are read as follows: Do not wait until the pointer comes to rest. Rather, while the pointer is still oscillating back and forth, read off its successive extreme positions. From these, guess where it will come to rest finally.

Use Aitken's δ^2 process to guess the final resting place of the pointer, given the following extreme positions:

14. 40

16. 11

14. 57

15. 96

14. 71

15. 83 .

Use any consecutive triple of extreme positions, such as 16. 11, 14. 57, 15. 96. As four such consecutive triples can be taken from the six readings given above, Aitken's δ^2 process can give four approximations for the final resting place of the pointer. How do they compare?

Chapter IX

MAXIMA AND MINIMA OF A FUNCTION OF ONE VARIABLE

0. Guide for the reader.

The theory of maxima and minima is thoroughly discussed in calculus texts. See, for instance Sections 3-5 and 3-6 of T-F; in the latter section there is a subsection entitled "Strategy for finding maxima and minima."

Calculators can be very helpful in finding maxima and minima. So this chapter should be read whenever you encounter maxima and minima. Most texts recommend plotting the function for which you seek the maxima and minima. Obviously the calculator can expedite this. After that, you have to find points at which the derivative is 0. Finding a point where a function (or its derivative) is zero is discussed in Chap. VII. Calculators can help very much there. So make sure you know the material in Chap. VII.

1. General techniques.

If you only wish a rough idea of the maxima and minima of a function, no theory is required. Just plot the function in enough detail. Suppose we wish relative maxima and minima of

(1.1) $$x + 1 - 2^x$$

for $-3 \leq x \leq 3$. We give some selected values in Table 1.1. "Obviously" the value

Table 1.1

x	$x + 1 - 2^x$
-3	-2.125
-2	-1.25
-1	-0.5
0	0
1	0
2	-1
3	-4

increases from $x = -3$ to $x = 0$ and decreases from $x = 1$ to $x = 3$. (Actually this is not obvious until one knows the derivative of (1. 1).) So clearly the absolute minimum is at $x = 3$, where (1. 1) has the value -4. There is a relative minimum at $x = -3$ where (1. 1) has the value -2.125.

To get a better idea of the maximum, take some values of x between 0 and 1, as in Table 1. 2. Here we first got the values at $x = 0.2$, 0.4, 0.6, and 0.8, after

Table 1. 2

x	$x + 1 - 2^x$
0. 2	0. 05130
0. 4	0. 08049
0. 6	0. 08428
0. 8	0. 05890
0. 42	0. 08207
0. 44	0. 08340
0. 46	0. 08446
0. 48	0. 08526
0. 50	0. 08579
0. 52	0. 08604
0. 54	0. 08603
0. 56	0. 08573
0. 58	0. 08515

which we knew we should get some values for x between 0. 4 and 0. 6. The largest value we see is 0. 08604, but the actual maximum value is likely just a shade larger.

So we say that the maximum value of (1. 1) is about 0. 0861. After all, how accurately do you need to know the maximum value? In most scientific situations, this is as accurate a value as is needed. Where does the maximum occur? Between $x = 0.52$ and $x = 0.54$, "obviously." (It is not obvious until one knows the derivative of (1.1).)

Of course you can do better if you can get the derivative. Suppose we need to do better. It is still useful to do first what we have done so far, so as to get the general idea. And indeed, most calculus texts recommend that one start by plotting the function.

So let us get the derivative, and do better. You will learn in due course that

(1. 2) $$\frac{d}{dx}(x + 1 - 2^x) = 1 - 2^x \ln 2 .$$

The HP-33E gives

(1. 3) $$\ln 2 \cong 0.69314\ 71806 .$$

For $-3 \le x \le 0$, we have $2^x \le 1$. So (1. 1) does indeed increase from $x = -3$ to $x = 0$. For $1 \le x \le 3$, we have $2^x \ge 2$. So (1. 1) does indeed decrease from $x = 1$ to $x = 3$. And indeed the maximum must occur between $x = 0$ and $x = 1$. Also our original determination of the two minima was right on the nose.

To find where the derivative is 0, we have to solve

$$2^x \ln 2 = 1.$$

By calculator, this gives

$$2^x \cong 1.4426\ 95041 .$$

Taking the ln of both sides gives

$$x \ \ln 2 \cong \ln (1.4426\ 95041)$$

$$\cong 0.36651\ 29207 .$$

So

$$x \cong \frac{0.36651\ \ 29207}{0.69314\ \ 71806} \cong 0.52876\ \ 63731 . \tag{1.4}$$

So this is where the maximum occurs. By calculator, the value of (1.1) at that point is about

$$0.0860\ \ 71332 .$$

As we said earlier, this is about 0.0861. Very seldom would one need a more accurate estimate of the maximum value than this.

The calculus text warns you to make various tests to find out if the x that makes the derivative equal to 0 gives a maximum or minimum. In view of Table 1.2, these tests are superfluous here.

Problem 1.1. Find approximately all local maxima and minima for

$$\frac{\ln x}{x} \tag{1.5}$$

for $1 \le x \le 10$. Do this simply by calculating enough values of (1.5).

Problem 1.2. Using the fact that

$$\frac{d}{dx} \ln x = \frac{1}{x} , \tag{1.6}$$

determine quite accurately all local maxima and minima of (1.5) for $1 \le x \le 10$.

Hint. Verify that for the number $e = e^1$, got by taking $x = 1$ and pressing the e^x key (or equivalent) on the calculator, the calculator gives $\ln e = 1$ to very high accuracy.

Problem 1. 3. Consider the trapezoid ABCD shown in Fig. 1. 1, with DA = 1,

Figure 1. 1

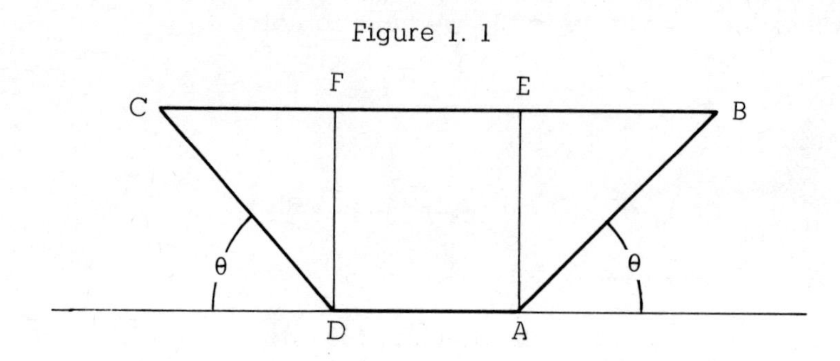

AB = CD = 2, and the angle θ unspecified. Draw perpendiculars AE and DF from A and D up to BC, as shown. Then

$$DF = AE = 2 \sin \theta$$

$$CF = EB = 2 \cos \theta \ .$$

The area of the trapezoid will be the sum of the area of the rectangle AEFD and of the two triangles ABE and DCF, and so

$$\text{Area} = 2 \sin \theta + 4 \sin \theta \cos \theta \ . \tag{1.7}$$

(a) Why is this formula for the area incorrect if $\theta > \cos^{-1}(-0.25)$?

(b) Show that if one is seeking to maximize the area, it suffices to restrict attention to $0 \leq \theta \leq \pi/2$, since if $\theta > \pi/2$, then the acute angle $\theta' = \pi - \theta$ gives a larger area.

In view of (b), it suffices to use (1. 7) for the area in trying to maximize the area, in spite of (a).

Differentiating the right side of (1. 7) and expressing everything in terms of $\cos \theta$ gives a quadratic formula in $\cos \theta$.

(c) Find approximately what value of θ gives the trapezoid a maximum area, and give the maximum area approximately.

(d) Be sure to show that your stipulated value of θ does indeed make the area a maximum.

2. Special techniques for polynomials.

What is special about polynomials is the fact, pointed out in Sect. 3 of Chap. II, that there are calculator programs which, for a given x, will calculate both the polynomial and its derivative for that x. See Program (3. 2) in Chap. II, or preferably Program II. 4 in the Program Appendix. From the values of the polynomial, you can draw its graph, which is the first step. The values of the derivative help you to find where the derivative is 0.

Let us find the relative maxima and minima of the polynomial, p, defined by

$$p(x) = x^4 - 4x^3 + 2x^2 - 4x + 1 \tag{2.1}$$

for $0 \le x \le 4$. We first make up Table 2.1, which gives some values of $p(x)$ and $p'(x)$ for x between 0 and 4.

Table 2.1

x	p(x)	p'(x)
0	1	-4
1	-4	-8
2	-15	-12
3	-20	8
4	17	76

It would seem likely, just from the values of $p(x)$ alone, that the absolute maximum occurs at 4, and is 17. Also that there is a relative maximum at $x = 0$, where $p(x) = 1$. This judgement is reinforced by the values recorded for the derivative, which verify that $x = 0$ and $x = 4$ are at least relative maxima. From the derivative values, we see that the derivative must be 0 for at least one point between 2 and 3. Accordingly, we make up Table 2.2, calculating first for $x = 2.2$, 2.4, 2.6, and 2.8, and then

Table 2.2

x	p(x)	p'(x)
2.2	-17.286 4	-10.688
2.4	-19.198 4	- 8.224
2.6	-20.486 4	- 4.416
2.8	-20.862 4	0.928
2.76	-20.875 27424	- 0.2728 96
2.77	-20.876 53759	0.0209 32
2.78	-20.871 04144	0.3190 08
2.79	-20.870 14319	0.6213 56

doing four more entries on the basis of what we learned from these. As the derivative is nearly 0 at $x = 2.77$, we can confidently assert that near there $p(x)$ has a minimum value of about -20.877.

This is certainly accurate enough for most purposes. Should we wish it more accurately, or wish to be sure about where $p(x)$ is increasing or decreasing, we can take the derivative. By (2.1)

$$p'(x) = 4(x^3 - 3x^2 + x - 1)\,. \tag{2.2}$$

In the course of solving Prob. 4.9 of Chap. VII, we get the factorization

$$p'(x) = 4(x - c)(x^2 + ax + b), \tag{2.3}$$

with

$$c \cong 2.7692\ 92354\ . \tag{2.4}$$

You can get all the zeros of $p'(x)$ by getting the zeros of $x^2 + ax + b$ by the quadratic formula. The result is that c is the only real zero of p'. For $x < c$, $p'(x)$ is negative and $p(x)$ is decreasing. For $x > c$, $p'(x)$ is positive and $p(x)$ is increasing. So the absolute minimum of $p(x)$ occurs at $x = c$. Substituting c into the right side of (2.1) gives

$$-20.876\ 54499, \tag{2.5}$$

which is a close approximation to the absolute minimum value of $p(x)$.

Problem 2.1. Find approximate local maxima and minima for

$$x^4 - x^2 - 0.4x \tag{2.6}$$

for $-3 \leq x \leq 3$.

Caution. Watch that you don't miss any relative maxima or minima.

Problem 2.2. Determine quite accurately the local maxima and minima of (2.6) for $-3 \leq x \leq 3$.

Hint. If you factor the derivative of (2.6), analogously to our (2.3), you shouldn't miss any zeros of the derivative.

Chapter X

NUMERICAL INTEGRATION

0. Guide for the reader.

The function F is an antiderivative or indefinite integral for the function f in case $F' = f$. Further, if also G is an antiderivative for f, then G and F differ only by a constant. These two facts are often expressed by writing

$$\int^{x} f(s)ds = F(x) + C \ , \tag{0.1}$$

where F is a particular antiderivative for f and C is a "constant of integration."

You will learn in the calculus the very important fact that every continuous function f has an antiderivative F. This can be put to use because the definite integral

$$\int_a^b f(x)\,dx \tag{0.2}$$

has the value expressed by

$$\int_a^b f(x)\,dx = F(b) - F(a) \ ; \tag{0.3}$$

this is the Fundamental Theorem of Integral Calculus. But if you wish to use (0.3) for the evaluation of (0.2), you must have a means of evaluating F at a and b.

This is possible if F can be given <u>in closed form.</u> That is, F can be expressed in terms of sines, cosines, exponentials, logarithms, and other functions for which you have keys on your calculator. In short, F can be keyed in on your calculator. In such a case, (0.3) furnishes a ready means to calculate the value of the definite integral

(0. 2). Then we say that (0. 3) gives the definite integral in closed form.

In calculus texts, attention is restricted almost entirely to functions f for which an antiderivative can be given in closed form. For that reason, much of the material in this chapter will not be needed in working the special sorts of problems given in your calculus course. However, after you have read the definition of a definite integral in your calculus text as a limit of sums of areas of rectangles, it will be advisable to read Sect. 1 in order to appreciate that, even with the help of a calculator, the definition of a definite integral does not give a practical means to calculate an approximation for its value. You most often need something like (0. 3), or one of the more efficient approximation methods given later in this chapter.

You will encounter some pretty complicated integrals later on, so complicated that it is easy to make a numerical mistake and get an answer that is way off. In Sect. 2 is given a simple way to get an approximate answer by calculator, which should let you know if you have made such a mistake. When you encounter such complicated integrals, you should read Sect. 2.

Are there functions f with an antiderivative F that cannot be given in closed form? Indeed there are. That does not mean that F is not a function. It only means that such an F is not useful when it comes to evaluating definite integrals. Specifically, if F cannot be given in closed form,

$$\int_a^b f(x)\,dx = F(b) - F(a)$$

does not give much clue as to how to calculate an approximation for

$$\int_a^b f(x)\,dx \; .$$

Some calculus texts give means for getting an approximation. If your calculus text gives the midpoint rule or the trapezoidal rule for this purpose, you should be sure to read Sect. 3 when you get to that point. If your calculus text gives Simpson's rule for this purpose, you should be sure to read Sect. 4 when you get to that point.

If your calculus text gives no means other than (0. 3) to get some sort of evaluation for the definite integral, the text will have to confine its definite integrals to those with an f such that (0. 3) gives the integral in closed form. As far as the calculus course is concerned, if you have such a text you have no need to read Sections 3 or 4.

Unfortunately, in engineering, physics, chemistry, and other quantitative sciences, as well as economics, one does encounter f's for which an antiderivative cannot be given in closed form, but for which a numerical approximation is needed for

$$\int_a^b f(x)\,dx \; .$$

In Sections 3 and 4, methods are given for getting a numerical approximation. If you plan to go on in one of the subjects above, you should prepare yourself by studying Sections 3 and 4 carefully. Now is the best time to get some help from your instructor or teaching assistant if you should encounter a difficulty in one of these sections.

One of the benefits of owning a calculator is the ability to cope with definite integrals that arise in applications and cannot be given in closed form.

One such is

$$\int_0^x \sin y^2 \, dy ,$$

which is extensively used in optics. Another is

$$\int_0^x e^{-y^2} \, dy ,$$

which plays a key role in statistics, and so appears in many applied situations. There are many others.

1. Definition of a definite integral; the rectangle rule.

By definition, the definite integral,

$$\int_a^b f(x)\, dx , \tag{1.1}$$

is the limit of sums of the form

$$f(c_1)h + f(c_2)h + \ldots + f(c_N)h$$

as $N \to \infty$. We write this as

$$\sum_{n=1}^{N} f(c_n)\, h . \tag{1.2}$$

These sums are called Riemann sums, and the N-th such sum is formed as follows. Divide the interval of integration, $a \leq x \leq b$, into N equal intervals by the $N+1$ points

$$x_n = a + nh, \qquad n = 0, 1, \ldots, N ,$$

with each subinterval of length

$$h = \frac{b-a}{N} .$$

For each $n = 1, 2, \ldots, N$, c_n is chosen somewhere in the n-th interval,

$$x_{n-1} \leq c_n \leq x_n .$$

You can visualize each summand, $f(c_n)h$, in (1. 2) as an area, the area of a rectangle of base $h = x_n - x_{n-1}$ and height $f(c_n)$. (This area turns out to be negative in case $f(c_n)$ is negative. You will have to get used to negative areas.)

In Fig. 1. 1, we have pictured one such sum, for the particular definite integral

$$\int_0^1 \frac{dx}{1 + x^2} . \tag{1. 3}$$

Figure 1. 1

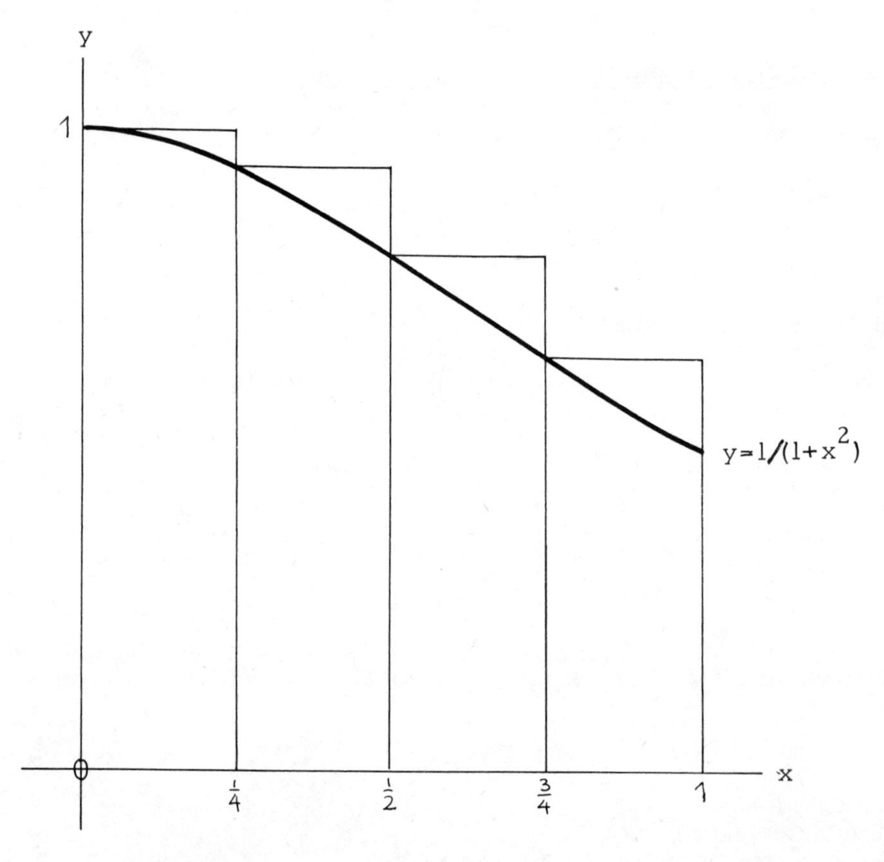

1. The rectangle rule

This integral gives the area under the curve

$$y = \frac{1}{1 + x^2} \tag{1.4}$$

between $x = 0$ and $x = 1$. In the figure, we have chosen each point c_n to be the left endpoint of the interval. Specifically,

$$c_n = x_{n-1} = a + (n - 1)h, \qquad n = 1, 2, \ldots, N.$$

The particular integral (1. 3) can be given in closed form. You should know that

$$\int^x \frac{dy}{1 + y^2} = \tan^{-1} x + C. \tag{1.5}$$

This says that an antiderivative F turns out to be $\tan^{-1}$, so that by (0. 3) and (1. 3) we get

$$\text{Area} = \int_0^1 \frac{dx}{1 + x^2} = \tan^{-1}(1) - \tan^{-1}(0) \tag{1.6}$$

$$= \frac{\pi}{4} = 0.78539\ 81633\ 97448\ldots\ .$$

But, now pretend that you do not know a closed form for (1. 3). You might then be tempted to use Riemann sums (1. 2) to approximate (1. 3). After all, (1. 3) is defined as the limit of such sums. Take, for example, the Riemann sum pictured in Fig. 1. 1. There we have four rectangles, each of width $1/4$, standing side by side from $x = 0$ to $x = 1$. Each has its upper left hand corner on the curve (1. 4), so that each has a height $1/(1 + x^2)$, for various values of x. The heights, widths, and areas are given in Table 1. 1. The sum of the areas of the rectangles is an approximation (not too good) to the

Table 1. 1

	1st rect.	2nd rect.	3rd rect.	4th rect.
height	$1/(1 + 0^2)$ $= 1.00000$	$1/(1 + (\frac{1}{4})^2)$ $= 0.94118$	$1/(1 + (\frac{1}{2})^2)$ $= 0.80000$	$1/(1 + (\frac{3}{4})^2)$ $= 0.64000$
width	$1/4 = 0.25$	$1/4 = 0.25$	$1/4 = 0.25$	$1/4 = 0.25$
approx. area	0. 25000	0. 23529	0. 20000	0. 16000

area we wish to determine. This sum of areas of rectangles is about 0.84529. As each rectangle sticks up above our curve, the sum should be too large, and by (1.6), it is.

If we use thinner rectangles, they would not stick up over the curve so badly, and the sum of the areas of the rectangles should be closer to the area under the curve. Let us take N rectangles, each of width $1/N$. The n-th rectangle would have its upper left hand corner at $x = (n-1)/N$, and so would be of height

$$\frac{1}{1+\left(\frac{n-1}{N}\right)^2}. \tag{1.7}$$

So the sum of the areas of the rectangles would be

$$\frac{1}{N}\left\{\frac{1}{1+0^2}+\frac{1}{1+\left(\frac{1}{N}\right)^2}+\frac{1}{1+\left(\frac{2}{N}\right)^2}+\dots+\frac{1}{1+\left(\frac{N-1}{N}\right)^2}\right\},$$

where we have factored the width, $1/N$, out of each of the areas. Using the Σ notation, as in (1.2) (see Sect. 4-5 of T-F), we can condense the above to

$$\frac{1}{N}\sum_{n=1}^{N}\frac{1}{1+\left(\frac{n-1}{N}\right)^2}. \tag{1.8}$$

As N goes to infinity, (1.8) will have the area under the curve as its limit. And the definite integral is defined as the same limit.

Let us try some calculations, to see how the limit is approached. You can program the calculator to take $n = 1, 2, \dots, N$ and add up the values (1.7), and finally divide by N. Some results are shown in Table 1.2, rounded to 8 digits after the decimal point. Indeed, the values are approaching the area, as given by (1.6). However,

Table 1.2

N	sums of areas of rectangles
10	0.80998 150
100	0.78789 400
1000	0.78564 812
10000	0.78542 316

even at $N = 10,000$, we still have only 4 decimals correct. And taking N as large as 10,000 leads to an extended calculation. You start the calculator going, and go to bed, and in the morning the sum is showing on the display.

So, approximating (1. 3) by using the definition is theoretically possible, but not really practical. Fortunately, there are better ways, which we will explain in Sections 3 and 4. In preparation, we learn some useful things by further discussion of the definition of the definite integral.

The key idea of the definite integral is to divide the interval from a to b into N equal intervals, each equal to

$$h = \frac{b-a}{N} . \tag{1. 9}$$

Each rectangle is of width h, and the n-th rectangle is of height

$$f(a + (n - 1)h) .$$

It is more convenient to number the rectangles from 0 to $N-1$, so that the n-th rectangle would be of height

$$f(a + nh) \tag{1. 10}$$

and of area

$$hf(a + nh) . \tag{1. 11}$$

So the sum of the areas of the rectangles is

$$h \sum_{n=0}^{N-1} f(a + nh) \tag{1. 12}$$

and the definite integral is defined as

$$\int_a^b f(x)dx = \lim_{N \to \infty} \sum_{n=0}^{N-1} f(a + nh) . \tag{1. 13}$$

As we saw in Table 1. 2, (1. 12) approaches its limit rather slowly for our example, so that for practical calculation it is not much help. Use of (1. 12) to approximate (1. 3) is called the <u>rectangle rule</u>, for obvious reasons, and (1. 12) is sometimes called $R(h)$.

One can write a formula for the difference between (1. 12) and its limit, namely

$$\frac{b-a}{2} f'(\xi) h , \tag{1. 14}$$

where ξ is some number between a and b. This is proved in Sect. 5. 3 of

"Elementary Numerical Analysis," second edition, by S. D. Conte and Carl de Boor, McGraw-Hill Book Co., 1972.

In other words, we have

$$\int_a^b f(x)dx = h\sum_{n=0}^{N-1} f(a + nh) + \frac{b-a}{2} f'(\xi)h . \tag{1.15}$$

Going back to (1.3) we have

$$f(x) = \frac{1}{1+x^2}$$

$$f'(x) = \frac{-2x}{(1+x^2)^2} .$$

If we knew what ξ is for some value of N, we could then use (1.15) to calculate the integral exactly. But, unfortunately, we have no clue what ξ is, except that it is between a and b.

In the case of (1.3), we know that ξ must be between 0 and 1. An elementary calculation ($f'(x)$ takes its minimum at $x = 1/\sqrt{3}$) tells us that

$$-0.65 < f'(\xi) < 0 . \tag{1.16}$$

So, taking $b = 1$ and $a = 0$ in the last term on the right of (1.15), we see that each entry in Table 1.2 should be more than the true value, but not by more than

$$0.325h = \frac{0.325}{N} . \tag{1.17}$$

Comparing with the true value, from (1.6), we see that this is indeed so. On the other hand, since the final term on the right of (1.15) could be as large as (1.17), that means that if we wish to be sure of approximating (1.5) to 10 decimals by (1.8), we will have to take N of the order of 10^{10}. This is quite hopeless with a hand held calculator, and would result in a very extended calculation even on a very fast big computer.

It is good there is something better, as we see below.

Incidentally, since we know the true value of (1.3) by (1.6), we can calculate approximately what $f'(\xi)$ is from (1.15) for each value of N in Table 1.2. These are given in Table 1.3.

By use of (1.15) this verifies that for this particular example, the error for the rectangle rule "behaves like h." By this, we mean the following. Define

2. Approximate check on complicated integrals

Table 1. 3

N	10	100	1000	10000
$f'(\xi)$	-0. 492	-0. 499	-0. 500	-0. 500

$$\text{error } (h) = \frac{\pi}{4} - \frac{1}{N} \sum_{n=0}^{N-1} \frac{1}{1 + (\frac{n}{N})^2} .$$

Then one has

$$\frac{\text{error } (h)}{h} \cong -0.25 .$$

Since we do not recommend the rectangle rule for approximating definite integrals, we have not set any problems for the student for this section.

2. Approximate check on a complicated definite integral.

Fancy methods are introduced in the calculus to help integrate complicated functions. These involve various substitutions, transformations, and adjustments. By the time you get the function integrated, you may have gone through several changes, in each of which there was a chance to get a factor wrong, or make some similar mistake. So it is worthwhile to have some sort of check, to see if your answer is anywhere near right.

Such a check should also be of help when your answer is correct but doesn't look at all like the answer in the book or your friend's answer. This situation is quite common, particularly with integrals involving trigonometric functions. For example, the three functions

$$(\cos x)^2, \quad -(\sin x)^2, \quad \text{and} \quad \frac{\cos 2x}{2}$$

are all antiderivatives for the same function. Again,

$$\sinh^{-1}(x/a) \quad \text{and} \quad \ln|x + \sqrt{a^2 + x^2}|$$

are antiderivatives for the same function.

To check if

$$(\cos x)^2 \quad \text{and} \quad \frac{\cos 2x}{2}$$

are antiderivatives for the same function, calculate their difference for several values of x. You should get the same difference for each value of x. For a definite integral, we can use some formulas which we will lift out of the next two sections.

We start with an easy one. In Prob. 13 at the end of Sect. 4-8 of T- F, we are asked to evaluate

$$\int_0^{\frac{\pi}{6}} \frac{\sin 2x}{\cos^2 2x}\, dx . \tag{2.1}$$

One thinks of taking

$$u = \cos 2x . \tag{2.2}$$

Since the derivative of cos is - sin, we have

$$\int \frac{-du}{u^2} = \frac{1}{u} = \frac{1}{\cos 2x} .$$

Putting in the limits 0 and $\frac{\pi}{6}$ gives

$$\frac{1}{\left(\frac{1}{2}\right)} - \frac{1}{1} = 1$$

as the answer.

Now let us try to check this by using the trapezoidal rule. We write (2. 1) as

$$\int_0^{\frac{\pi}{6}} f(x)\, dx ,$$

with

$$f(x) = \frac{\sin 2x}{(\cos 2x)^2} . \tag{2.3}$$

The two simplest trapezoidal rules, which will be got by taking $N = 1$ and $N = 2$ in (3. 5) or (3. 6) ahead, are

$$\int_a^b f(x)dx \cong \frac{b-a}{2} \{ f(a) + f(b) \} \tag{2.4}$$

$$\int_a^b f(x)dx \cong \frac{b-a}{4} \{ f(a) + 2f(\tfrac{a+b}{2}) + f(b) \} . \tag{2.5}$$

2. Approximate check on complicated integrals

We have for (2. 1)

$$b - a = \frac{\pi}{6} \cong 0.52359\ 87756 .$$

The simplest trapezoidal rule approximation is, by (2. 4),

$$\frac{\pi}{6}\left\{\frac{1}{2} f(0) + \frac{1}{2} f\left(\frac{\pi}{6}\right)\right\} \cong \frac{\pi}{6}\left\{\frac{1}{2}(0) + \frac{1}{2}(3.4641\ 01615)\right\}$$

$$\cong 0.90689\ 96821 .$$

As this agrees within less than 10 %, we feel encouraged. However, maybe a trapezoidal rule on only two points could be considerably in error. Let us try one on three points.

Call T_2 the approximation we got above. For three points, we would get, by (2. 5),

$$T_3 = \frac{1}{2}\frac{\pi}{6}\left\{\frac{1}{2} f(0) + f\left(\frac{\pi}{12}\right) + \frac{1}{2} f\left(\frac{\pi}{6}\right)\right\}$$

$$= \frac{1}{2} T_2 + \frac{1}{2}\frac{\pi}{6} f\left(\frac{\pi}{12}\right)$$

$$\cong \frac{1}{2} T_2 + \frac{1}{2}\frac{\pi}{6}(0.66666\ 66667)$$

$$\cong 0.62798\ 27663 .$$

According to (3. 7) ahead, T_3 should be about four times closer to the true integral than T_2, suggesting that the true integral is approximately

$$\frac{4T_3 - T_2}{3} \cong 0.53501\ 04610 . \tag{2. 6}$$

According to (4. 5) ahead, the simplest form of Simpson's rule is

$$\frac{b-a}{6}\left\{f(a) + 4f\left(\frac{a+b}{2}\right) + f(b)\right\} . \tag{2. 7}$$

This gives the same value as in (2. 6).

It does look as though perhaps we made a mistake in our original integration, as indeed we did. If we take u as in (2. 2), then

$$du = (-\sin 2x)(2\,dx) .$$

So we should have

$$\int \frac{\sin 2x}{(\cos 2x)^2}\,dx = \frac{-1}{2}\int \frac{(-\sin 2x)(2\,dx)}{(\cos 2x)^2}$$

$$= \frac{-1}{2}\int u^{-2}\,du$$

$$= \frac{1}{2}u^{-1}$$

$$= \frac{1}{2\cos 2x} .$$

So

$$\int_0^{\frac{\pi}{6}} \frac{\sin 2x}{(\cos 2x)^2}\,dx = \left[\frac{1}{2\cos 2x}\right]_0^{\frac{\pi}{6}} = \frac{1}{2\left(\frac{1}{2}\right)} - \frac{1}{2(1)} = \frac{1}{2} .$$

The value that we got above by Simpson's rule, namely

$$0.53501\ 04610 ,$$

corroborates our value of $1/2$ very well.

Of course, our numerical approximation could possibly still be a long way from the true answer. Certainly, if there are appreciable irregularities of f in the interval of integration, we should use more intervals for our trapezoidal rule or Simpson's rule. However, the function we were considering, given by (2.3), is very well behaved between 0 and $\pi/6$. So the close agreement between our Simpson's rule approximation and the answer of $1/2$ is a fairly good indication that the latter is correct (after our unfortunate blunder at first).

Problem 2.1. Evaluate each of the following definite integrals by

$$\int_a^b f(x)\,dx = F(b) - F(a) ,$$

where F is a suitable antiderivative of f in closed form. Then check whether you made a mistake, by approximating the integral by the simplest Simpson's rule, namely (2.7).

(a) $\int_0^2 \sqrt{4x+1}\ dx$

(b) $\int_{\frac{1}{\sqrt{3}}}^{1} \frac{dx}{x\sqrt{4x^2-1}}$

(c) $\int_1^3 \frac{dx}{x^2-2x+5}$

(d) $\int_0^{\frac{\pi}{2}} \frac{dx}{2+\cos x}$

(e) $\int_3^8 \frac{(t+2)dt}{t\sqrt{t+1}}$.

Remark. These integrals occur in T- F as problems at the ends of Sections, as follows respectively: Prob. 9, Sect. 4-8; Prob. 21, Sect. 6-3; Prob. 1, Sect. 7-5; Prob. 4, Sect. 7-8; Prob. 9, Sect. 7-9.

Problem 2. 2. Do the same as in Prob. 2. 1 with the integral

$$\frac{32}{\sqrt{64.\,32}} \int_{60}^{70} \frac{(400-(H-60)^2)}{\sqrt{H}}\ dH$$

that will result from (2. 16) in Chap. XV if we take $a = \pi/16$.

Hint. To get an antiderivative, substitute $H = x^2$.

3. The midpoint rule and the trapezoidal rule.

In defining the definite integral, we divided the interval from a to b into N equal intervals, each of length h. We took rectangles of width h, filling up the whole area, except just close to the curve. We took the heights of the rectangles to be the values of f at the left ends of the intervals. Let us instead take the heights of the rectangles to be the values of f at the midpoints of the intervals. This is illustrated for the case N = 4 in Fig. 3. 1. It is clear that rectangles of these heights approximate much more closely the area of the curve above the interval than when the left hand upper corner was on the curve.

Whereas for the rectangle rule we took (1. 12) as the approximation for the definite integral, we are now proposing to take

$$M(h) = h\sum_{n=0}^{N-1} f(a+(n+\tfrac{1}{2})h) \ . \tag{3. 1}$$

Figure 3. 1

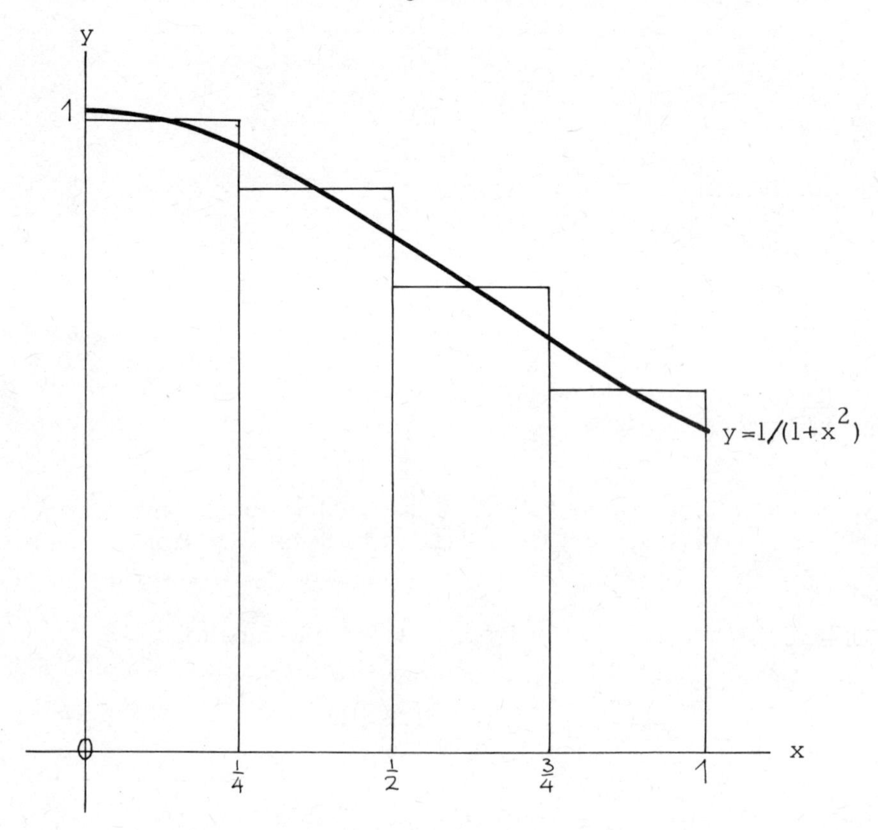

If we let $N \to \infty$, this will have the definite integral as its limit.

Use of this to approximate a definite integral is called the <u>midpoint rule</u>.

One can write a formula for how far off $M(h)$ is from its limit, namely

$$\frac{b-a}{24} f''(\xi) h^2 , \tag{3.2}$$

where ξ is some number between a and b. This is proved in the same reference we gave for (1. 14). So we have

$$\int_a^b f(x)dx = M(h) + \frac{b-a}{24} f''(\xi) h^2 . \tag{3.3}$$

In Table 3. 1 we give a couple of instances of using $M(h)$ to approximate (1. 3),

Table 3.1

N	M(h)	M(h) plus correction
10	0.78560 64964	0.78539 81631
100	0.78540 02467	0.78539 81634

namely

$$\int_0^1 \frac{dx}{1+x^2} .$$

We clearly have a tremendous improvement over Table 1.2, which was calculated using the rectangle rule. We will later suggest a correction.

For $N = 100$, the calculation of $M(h)$ for (1.3) took only a very few minutes. Not that one should try it unless one has a programmable calculator. The program is very easy to write. For $N = 100$, it consists of the following operations. We will let n run down from 99 to 0 instead of up from 0 to 99, because many calculators have special commands to stop when a quantity gets to 0. We repeat the following steps for each value of n:

1. Start with $n + 1$ stored in register zero, h stored in register one, and a partial summation of the Σ in (3.1) stored in register two.

2. See if $n + 1 = 0$. If so, stop. Otherwise subtract 1 and store back in register zero.

3. Now you have n. Using the h stored in register one, calculate

$$\frac{1}{1 + ((n + \frac{1}{2})h)^2} .$$

4. Add the above to what is in register two. Most calculators have a special instruction for this.

5. Go back to the first step.

When you finish, the value of $M(h)$ will be stored in register two.

We will set up a problem involving the approximation of

$$\int_1^2 \frac{dx}{x} . \tag{3.4}$$

It will be shown in due course, if you have not already learned it, that the value of (3. 4) is $\ln 2$, for which your calculator will give a fairly accurate value.

If you have a nonprogrammable calculator, do (i) below only for $N = 1, 2, 4,$ and 8. Incidentally, even nonprogrammable calculators usually have a key that will add what is in the display to what is stored in a particular memory register. This greatly speeds up the calculation of a Σ, such as occurs in (3. 1).

Before attempting the problem, review the latter part of Sect. 1.

Problem 3. 1. (In five parts.)

(i) Use the midpoint rule to approximate (3. 4) for $N = 1, 2, 4, 8, 16$, and compare with the calculator approximation for $\ln 2$. Carry your calculations to the maximum precision possible, and save the results for future use in Sect. 4.

(ii) Should the midpoint rule give a value larger or smaller than $\ln 2$?

(iii) Verify that the error for the midpoint rule "behaves like h^2." That is, find a constant, K, such that with the error given by

$$\text{error}\ (h) = \ln 2 - h \sum_{n=0}^{N-1} \frac{1}{1 + (n + \frac{1}{2})h},$$

one has

$$\frac{\text{error}\ (h)}{h^2} \cong K .$$

(iv) From your value of K in (iii), estimate what value of N one should take to give 5 decimal accuracy by the midpoint rule; that is to make

$$|\text{error}\ (h)| \le 5 \times 10^{-6} .$$

(v) Estimate what value of N would be required to give 10 decimal accuracy by the midpoint rule.

Problem 3. 2. For the definite integral (3. 4), what is the largest possible value that $|f''(\xi)|$ could have, for the $f''(\xi)$ of (3. 3)? Using this, how large would you have to take N in order for the midpoint rule for (3. 4) to be guaranteed to give 5 decimal accuracy (that is, to guarantee that the final term on the right of (3. 3) is no greater than 5×10^{-6} in absolute value)? How large would you have to take N to guarantee 10 decimal accuracy?

It is possible to write the error (3. 2) in the midpoint rule in the alternate form

$$\frac{h^2}{24} \int_a^b f''(x)\,dx - \frac{7h^4}{5760} \int_a^b f^{iv}(x)\,dx + \dots .$$

For "small" h, the error should be mostly given by the first term, namely

$$\frac{h^2}{24}\int_a^b f''(x)\,dx = \frac{h^2}{24}[f'(b) - f'(a)] .$$

But this being so, computing and then adding this first term to the midpoint rule should give us a more accurate approximation.

For example, the integrand of the integral (1. 3)

$$\int_0^1 \frac{dx}{1+x^2}$$

has its derivative given by

$$f'(x) = -\frac{2x}{(1+x^2)^2} .$$

Consequently, the correction term becomes in this case the expression

$$\frac{h^2}{24}[f'(b) - f'(a)] = \frac{h^2}{24}\frac{-2}{(1+1)^2} = -h^2/48 .$$

We have added this term to the midpoint rule approximations shown in Table 3. 1, with the resulting approximation also shown in that table. The increase in accuracy in this example is dramatic.

Problem 3. 3. For the same values of N as in Prob. 3. 1, add the correction term

$$\frac{h^2}{24}[f'(b) - f'(a)]$$

to the results of Prob. 3. 1, and compare with the calculator approximation for $\ln 2$. Verify that the error for the corrected results "behave like h^4."

Some calculus texts give something called the trapezoidal rule for approximating definite integrals. See (1b) in Sect. 4-9 of T- F. It is derived as follows. Divide the interval from a to b into N equal intervals, each of length h. Instead of putting a rectangle on top of each interval, we put a trapezoid. This is shown for $N = 4$ in Fig. 3. 2. The trapezoids mostly fit the curve very well. As shown in Fig. 3. 2, each trapezoid has each of the two top corners on the curve. So the n-th trapezoid (we are numbering from 0 to $N-1$) has vertical sides equal to $f(a + nh)$ and $f(a+ (n+1)h)$. The area of such a trapezoid is

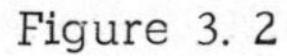

Figure 3. 2

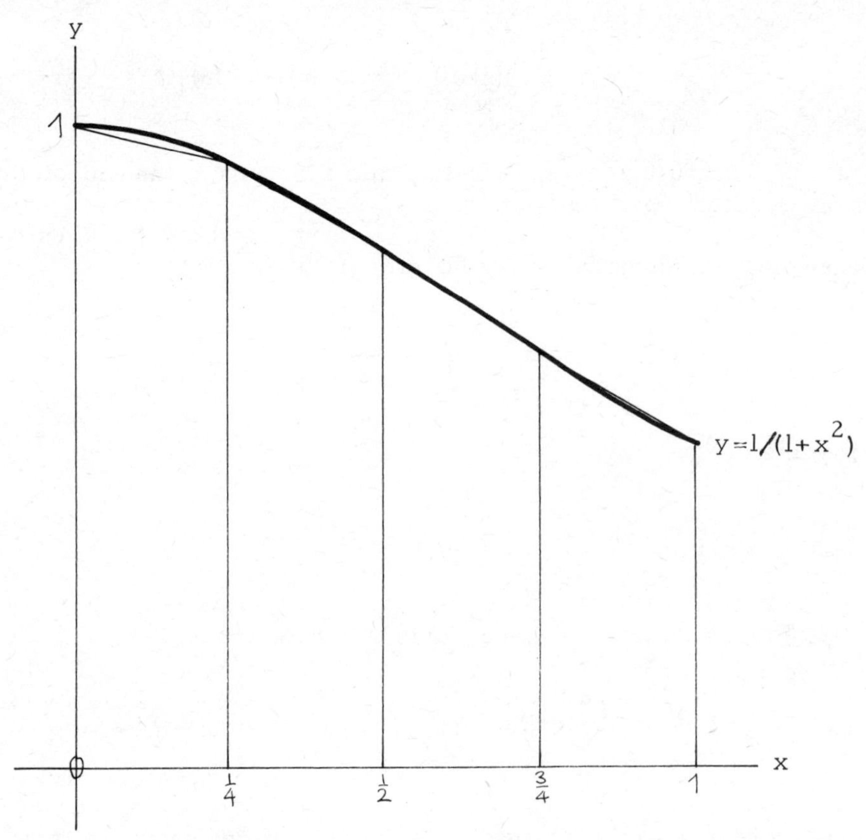

$$\frac{1}{2}h(f(a + nh) + f(a + (n + 1)h)) .$$

The sum of these areas is

$$T(h) = \frac{1}{2}h\{f(a) + 2f(a + h) + 2f(a + 2h) + \ldots + 2f(b - 2h) + 2f(b - h) + f(b)\} . \tag{3. 5}$$

With the Σ notation, one can write (3. 5) as

$$T(h) = \frac{1}{2}h\{f(a) + f(b)\} + h\sum_{n=1}^{N-1} f(a + nh) . \tag{3. 6}$$

If we let $N \to \infty$, this will have the definite integral as its limit.

Use of this to approximate a definite integral is called the trapezoidal rule (or sometimes the trapezoid rule).

One can write a formula for how far off $T(h)$ is from its limit, namely

$$-\frac{b-a}{12} f''(\xi)h^2 , \tag{3.7}$$

where ξ is some number between a and b. This is proved in the same reference we gave for (1.14), and is stated as (3) in Sect. 4-9 of T-F.

So we have

$$\int_a^b f(x)dx = T(h) - \frac{b-a}{12} f''(\xi)h^2 . \tag{3.8}$$

In Table 3.2 we give a couple of instances of using $T(h)$ to approximate (1.3). The calculations for Table 3.2 proceed very similarly to those for Table 3.1.

Table 3.2

N	T (h)	T(h) plus correction
10	0.78498 14974	0.78539 81641
100	0.78539 39969	0.78539 81636

If you have a nonprogrammable calculator, do (i) below only for $N = 1,2,4$, and 8.

Problem 3.4. (In five parts.)

(i) Use the trapezoidal rule to approximate (3.4) for $N = 1, 2, 4, 8, 16$, and compare with the calculator approximation for $\ln 2$. Carry your calculations to the maximum precision possible, and save the results for future use in Sect. 4.

(ii) Should the trapezoidal rule give a value larger or smaller than $\ln 2$?

(iii) Verify that the error for the trapezoidal rule "behaves like h^2" by getting a K such that the error is approximately Kh^2.

(iv) Using this K, estimate what value of N would be required to give 5 decimal accuracy by the trapezoidal rule.

(v) Estimate what value of N would be required to give 10 decimal accuracy by the trapezoidal rule.

It is possible to write the error (3. 7) in the trapezoid rule in the useful alternate form

$$-\frac{h^2}{12}\int_a^b f''(x)\,dx + \frac{h^4}{720}\int_a^b f^{iv}(x)\,dx - \ldots .$$

But then, as was pointed out earlier in connection with such an expansion for the error in the midpoint rule, there is a gain in accuracy available for sufficiently small h by evaluating the first term

$$-\frac{h^2}{12}\int_a^b f''(x)dx = -\frac{h^2}{12}[f'(b) - f'(a)]$$

and adding this number to the trapezoid rule.

For example, for the integrand of the integral (1. 3), this gives the correction term

$$-\frac{h^2}{12}[f'(b) - f'(a)] = -\frac{h^2}{12}\frac{-2}{(1+1)^2} = h^2/24 .$$

We have added this term to the trapezoid rule approximations given in Table 3. 2, with the resulting approximation also shown in that table. Just as was the case with the midpoint rule, these resulting "corrected" approximations are a great improvement in this example.

Problem 3. 5. For the same values of N as in Prob. 3. 4, add the correction term

$$-\frac{h^2}{12}[f'(b) - f'(a)]$$

to the results of Prob. 3. 4, and compare with the calculator approximation for $\ln 2$. Verify that the error for the corrected results "behaves like h^4."

4. Simpson's rule.

Some calculus texts give something called Simpson's rule for approximating definite integrals. See (5) in Sect. 4-9 of T-F.

To get Simpson's rule, we proceed as follows. Take $M = 2N$, and divide the interval from a to b into M equal parts, each of length $h = (b-a)/M$. We now take the parts in pairs. We can do this, since M is an even number, $M = 2N$. In fact we will have N pairs. Let a pair go from x_r to $x_r + 2h$. We erect vertical lines at x_r and $x_r + 2h$. On top we put a parabola which passes through the three points on the curve where $x = x_r$, $x_r + h$, and $x_r + 2h$, respectively. The theory is that, since the parabola is curved, it fits the true curve better than the straight lines (tops of rectangles or trapezoids) that we have been putting on top. So we take the area under the parabola. We do this for each of the N pairs, and take the sum of the areas as an approximation for the true area under the curve.

To find the area under the parabola (for each parabola) we proceed as follows. Let

$$y = Ax^2 + Bx + C = g(x)$$

be the equation of the parabola. That is, if $y = f(x)$ is the curve we are trying to integrate, we have $f(x) = g(x)$ for $x =$ each of the values x_r, $x_r + h$, and $x_r + 2h$, since we are supposing that the parabola was chosen to coincide with the original curve at these three values of x.

Now, for any ξ whatsoever, we have $g''(\xi) = 2A$. So, taking $a = x_r, b = x_r + 2h$, $2h$ for h, and g for f gives exactly

$$\int_{x_r}^{x_r+2h} g(x)dx = M(2h) + \frac{2h}{24}(2A)(2h)^2$$

from (3. 3) and exactly

$$\int_{x_r}^{x_r+2h} g(x)dx = T(2h) - \frac{2h}{12}(2A)(2h)^2$$

from (3. 8). Now multiply the first equation by 2 and add the second equation. This gives

$$3\int_{x_r}^{x_r+2h} g(x)dx = 2M(2h) + T(2h) .$$

So

$$\int_{x_r}^{x_r+2h} g(x)dx = \frac{2}{3}M(2h) + \frac{1}{3}T(2h) . \tag{4. 1}$$

Here, of course, we have to be using the values of $g(x)$ in evaluating $M(2h)$ and $T(2h)$, since we are finding the area under the parabola. But these values are taken at points where $g(x) = f(x)$. So, on the right side of (4. 1), we may take the $M(2h)$ and $T(2h)$ as referring to $f(x)$.

We get (4. 1) for the area under each parabola, with the $M(2h)$ and $T(2h)$ referring to $f(x)$. Adding them together suggests that we define the Simpson's rule approximation, $S(h)$, for the definite integral by

(4. 2) $$S(h) = \frac{2}{3} M(2h) + \frac{1}{3} T(2h) .$$

If we let $h \to 0$, $S(h)$ will have the definite integral as its limit.

One can write a formula for how far $S(h)$ is from its limit, namely

(4. 3) $$-\frac{b-a}{180} f^{iv}(\xi) h^4 ,$$

where ξ is some number between a and b. This is proved in the same reference we gave for (1. 14). So we have

(4. 4) $$\int_a^b f(x)dx = S(h) - \frac{b-a}{180} f^{iv}(\xi) h^4 .$$

In Table 4. 1 we give a couple of instances of using $S(h)$ to approximate (1. 3).

Table 4. 1

M	N	S(h)
20	10	0. 78539 81633
200	100	0. 78539 81633

Comparing with (1. 6), we see that, even at $M = 20$, we are off by only a single unit in the 10-th decimal place; this is a remarkably small roundoff error, considering the amount of calculation it took to get to $S(h)$.

Needless to say, we calculated $S(h)$ by (4. 2), using the values of $M(2h)$ and $T(2h)$ from Tables 3. 1 and 3. 2 respectively. So the calculation of $S(h)$ presents no particular difficulty. You just calculate $M(2h)$ and $T(2h)$ and use (4. 2).

Problem 4. 1. (In five parts.)

(i) Use Simpson's rule to approximate (3. 4) for $M = 2, 4, 8, 16$, and compare with the calculator approximation for $\ln 2$. (Hint. Use (4. 2), taking $M(2h)$ and $T(2h)$ from your answers for Problems 3. 1 and 3. 3. Recall that $M = 2N$.)

(ii) Should Simpson's rule give a value larger or smaller than $\ln 2$?

(iii) Verify that the error for Simpson's rule "behaves like h^4" by getting a K such that the error is approximately Kh^4.

(iv) Using this K, estimate what value of M would be required to give 5 decimal accuracy by Simpson's rule.

(v) Estimate what value of M would be required to give 10 decimal accuracy by Simpson's rule.

Problem 4. 2. For the definite integral (3. 4), what is the largest possible value that $|f^{iv}(\xi)|$ could have, for the $f^{iv}(\xi)$ of (4. 4)? Using this, how large would you have to take M in order for Simpson's rule for (3. 4) to be guaranteed to give 5 decimal accuracy? How large would you have to take M to guarantee 10 decimal accuracy?

Problem 4. 3. Explain why Simpson's rule will give the exact value of the definite integral if f is a polynomial of degree 3 or less.

Suppose f is a polynomial of degree 3 or less. By Prob. 4. 3, Simpson's rule will give the exact value of the definite integral, no matter what M is. (Don't forget that M must be even.) Take $M = 2$. For the midpoint rule, we need to know only $f((a + b)/2)$. For the trapezoidal rule, we need to know only $f(a)$ and $f(b)$. From these, we get $S(h)$ by (4. 2). As f is a polynomial, the required values of f can quickly be calculated by Horner's method (see Sect. 1 of Chap. II).

Problem 4. 4. Calculate a highly precise approximation for

$$\int_{-2}^{3} (x^3 - 5x^2 - 3x + 7)\, dx$$

by Simpson's rule. Check by (0. 3).

It is possible to give also for the error (4. 3) in Simpson's rule an alternate expression (as we did earlier for the error in the midpoint rule and in the trapezoid rule), namely

$$-\frac{h^4}{180}\int_a^b f^{iv}(x)dx + c_6 h^6 \int_a^b f^{vi}(x)dx + \ldots\ .$$

But, while it is possible, in principle, to calculate the first term

$$-\frac{h^4}{180}\int_a^b f^{iv}(x)\, dx = -\frac{h^4}{180}[f'''(b) - f'''(a)]$$

in this expansion and add it to the Simpson's rule estimate, it requires calculating three derivatives. If the accuracy of a particular Simpson's rule estimate is not satisfactory, it is usually much easier to rerun Simpson's rule for a somewhat larger N or M, particularly if one has a programmable calculator with a canned Simpson's rule program.

In Sect. 4-9, T-F derive Simpson's rule by the same basic scheme. However, they work out formulas for what the A, B, and C of the equation of the parabola must be. Then they get

$$\int_{x_r}^{x_r+2h} g(x)\,dx$$

by actually integrating. They give

$$(4.5)\qquad \frac{h}{3}\{f(a) + 4f(a+h) + 2f(a+2h) + 4f(a+3h)$$

$$+ 2f(a+4h) + \ldots + 2f(b-4h) + 4f(b-3h)$$

$$+ 2f(b-2h) + 4f(b-h) + f(b)\}$$

as the formula for Simpson's rule. One could get the same formula from our definition (4. 2) by putting in the definitions of $M(2h)$ and $T(2h)$.

In (6) of Sect. 4-9, T-F, give the same formula for the error of (4. 5) that we have given in (4. 4).

To evaluate the Simpson formula (4. 5) directly is not all that hard, but one has to be careful to take care of a number of details. One has to get $f(a)$ at one end and $f(b)$ at the other, without 2 or 4 as coefficients. In the middle, one has to alternate the coefficients 2 and 4, starting with 4. To accomplish this, start with -1 in register 0. Then, every time you need the next coefficient, execute

RPN
|
RPN

3, [RCL 0], [CHS], [STO 0], [+] .

AE
|
AE

3, [+], [RCL 0], [+/-], [STO 0], [=] .

The better calculators have a lot of special programs already written for them, which one can buy. Usually one of them is a program to calculate (4. 5). The different calculator manufacturers have a variety of mechanisms for arranging that you can use the program conveniently on their calculators. You will have to read their manuals. If the program is labelled only "Numerical Integration", you should check to see if it uses Simpson's rule or one of the others. Also, if it depends on N, is that what we have called M, or what we have called N ?

Chapter XI

VECTOR MANIPULATION

0. Guide for the reader.

After you have learned about polar coordinates, you can benefit from reading Sections 1, 2, and 3. You should read these sections when vectors in the plane are covered in your course. Sect. 2 describes a game which may help you to become comfortable with vectors and, in any case, is fun to play.

Sections 4, 5, and 6 deal with 3-dimensional vectors, or "vectors in space", as some textbooks call them. You should read these sections carefully when this topic is covered in your course.

1. Length and direction of a vector in the plane.

We follow T-F and write a vector $\vec{A}$ in the (x, y)-plane as

$$A = a_1 \vec{i} + a_2 \vec{j} , \tag{1.1}$$

with $\vec{i}$ the unit vector in the positive direction of the x-axis, and $\vec{j}$ the unit vector in the positive direction of the y-axis.

We assume that you have learned from your textbook how to add vectors (componentwise), and how to multiply a vector by a scalar (again componentwise). You should also be aware of the fact that each vector $\vec{A}$ has a length $|\vec{A}|$ which is calculated from its components by

$$|\vec{A}| = \sqrt{a_1^2 + a_2^2} .$$

Further, if $|\vec{A}| \neq 0$, then the direction of $\vec{A}$ is, by definition, the unit vector

$$\text{dir}(\vec{A}) = \frac{1}{|\vec{A}|} \vec{A} = \frac{a_1}{|\vec{A}|} \vec{i} + \frac{a_2}{|\vec{A}|} \vec{j} .$$

For example, the vector $\vec{i}$ is its own direction, while the vector $\vec{i} + \vec{j}$ has direction

$$\frac{1}{\sqrt{2}}\vec{i} + \frac{1}{\sqrt{2}}\vec{j}.$$

The direction of $\vec{A}$, being a unit vector, can be visualized as an arrow from the origin to a point on the unit circle. Thus

$$\text{dir}(\vec{A}) = \cos\theta_A \vec{i} + \sin\theta_A \vec{j}$$

for a unique θ_A with $-180^\circ = -\pi < \theta_A \leq \pi = 180^\circ$. See Fig. 1. 1, where θ_A is negative. This θ_A is the angle between the positive x-axis and the vector $\vec{A}$, and is the direction angle of $\vec{A}$. For this chapter only, it is permissible to take angles in degrees.

Figure 1. 1

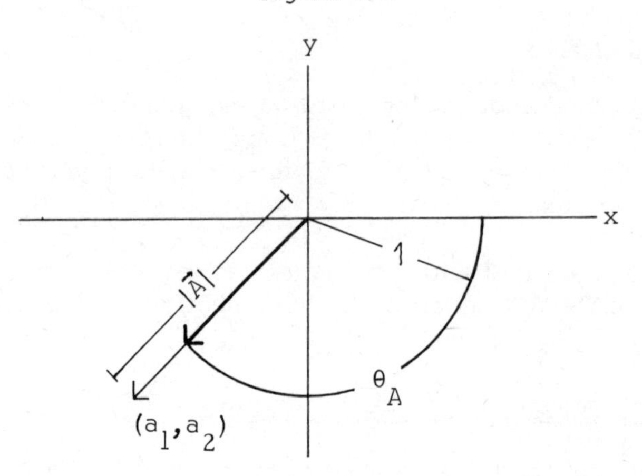

Length and direction (or direction angle) of a vector $\vec{A}$ provide an alternative way to describe the vector $\vec{A}$ given by (1. 1):

(1. 2) $$\vec{A} = |\vec{A}|\text{dir}(\vec{A}) = (|\vec{A}|\cos\theta_A)\vec{i} + (|\vec{A}|\sin\theta_A)\vec{j}.$$

This formula tells how to calculate the components of $\vec{A}$ from its length $|\vec{A}|$ and its direction angle θ_A. It is just a bit trickier to calculate length and direction angle from the components a_1 and a_2 of $\vec{A}$. We mentioned already that $|\vec{A}| = \sqrt{a_1^2 + a_2^2}$. For the direction angle, we must have

$$\cos\theta_A = a_1/|\vec{A}|, \quad \text{and} \quad \sin\theta_A = a_2/|\vec{A}|.$$

This seems to indicate that

$$\theta_A = \cos^{-1}(a_1/|\vec{A}|) \quad \text{and} \quad \theta_A = \sin^{-1}(a_2/|\vec{A}|).$$

But, unthinking use of the $\cos^{-1}$ or $\sin^{-1}$ key on your calculator is apt to bring you grief in this matter.

For example, the vector depicted in Fig. 1. 1,

$$\vec{A} = -\vec{i} - \vec{j} ,$$

clearly points southwest, and hence makes an angle of -135° (or 225°) with the x-axis. Also, $|\vec{A}| = \sqrt{(-1)^2 + (-1)^2} = \sqrt{2}$. But, using the $\cos^{-1}$ key on your calculator, you will find

$$\theta_A = \cos^{-1}(a_1/|\vec{A}|) = \cos^{-1}(-1/\sqrt{2}) = 135^{\circ} .$$

Perhaps, we fare better by using the equation $\sin\theta_A = a_2/|\vec{A}|$ instead? Try it. You will get

$$\theta_A = \sin^{-1}(a_2/|\vec{A}|) = \sin^{-1}(-1/\sqrt{2}) = -45^{\circ} ,$$

which isn't right, either.

Figure 1. 2

-135° -45° 135° θ_A

sign of a_1, a_2 (-,-) (+,-) (+,+) (-,+)

The difficulty stems from the fact that the $\cos^{-1}$ key implements the inverse function for $\cos\theta$, with θ restricted to the interval $0 \leq \theta \leq 180^{\circ} = \pi$. This means that you will find θ_A correctly from evaluating the expression $\cos^{-1}(a_1/|\vec{A}|)$ as long as θ_A lies between 0 and $\pi = 180^{\circ}$, i. e., as long as $\vec{A}$ points upwards or horizontal; what is the same thing, as long as $a_2 \geq 0$. If, on the other hand, $a_2 < 0$, i. e., if $\vec{A}$ points downwards, then $-180^{\circ} = -\pi < \theta_A < 0$, and then

$$\cos^{-1}(\cos\theta_A) = -\theta_A .$$

This then tells us to calculate $|\vec{A}|$ and θ_A from a_1 and a_2 as follows:

$$(1.3)\qquad |\vec{A}| = \sqrt{a_1^2 + a_2^2}$$

$$\theta_A = \begin{cases} \cos^{-1}(a_1/|\vec{A}|), & \text{if } a_2 \geq 0 \\ -\cos^{-1}(a_1/|\vec{A}|), & \text{if } a_2 < 0 \end{cases}.$$

Problem 1.1. Determine the length and direction angle for each of the following vectors:

(a) $\vec{i}$; (b) $2\vec{i} - 3\vec{j}$; (c) $-10\vec{i} + \vec{j}$; (d) $-\vec{i} - 3\vec{j}$.

Note. The calculation of $|\vec{A}|$ and θ_A from the components a_1 and a_2 of $\vec{A}$ amounts to finding the polar coordinates $(r, \theta) = (|\vec{A}|, \theta_A)$, with $r \geq 0$, of the point with cartesian coordinates (a_1, a_2). Fancier calculators have a special key for this conversion from cartesian or Rectangular coordinates to Polar coordinates (and one for the reverse). If you are blessed with such a calculator, find out how to use these special keys and get some exercise with them, for example by doing the problems in this section and the next two.

Problem 1.2. Describe how to use the $\sin^{-1}$ key on your calculator correctly to find θ_A from a_1 and a_2. (Hint. See Fig. 1.2).

Problem 1.3. An observer is looking for a tower which is supposedly 8.3 km west and 3.4 km south of his position. In which direction should he be looking?

Problem 1.4. A boat is crossing a river at 6.3 m/h, heading ESE. The river current carries the boat in a NNE direction at 1.7 m/h. What is the total motion of the boat (with respect to the land), i.e., what is its speed and its course? (Recall that SE is halfway between South and East, and that ESE is halfway between East and SE.)

Remark. The two motions mentioned can each be described by a vector whose length is the speed and whose direction is the direction of the motion. The combined motion is the vector sum of the individual motions. Since vectors are added componentwise, this problem then requires you to convert from polar to rectangular, add, and then convert back from rectangular to polar.

Problem 1.5. Determine the ground speed and the course of an airplane which flies a heading of 168° West from North at 350 knots in a 30 knot west wind (i.e., a wind blowing from the west).

Remark. You reach the direction 168° West from North by going 168° counterclockwise from due North.

The direction angle is handy to have around when it comes to rotating coordinates. Suppose you wish to rotate your coordinate system counterclockwise by an angle α. If the point A has coordinates (a_1, a_2) in the old coordinate system, what will its coordinates be in the new?

Since the x-axis is being rotated counterclockwise by an angle α, the direction angle θ_A for the vector $\vec{A}$ from the origin to the point A is decreased by α in the new coordinates. Thus, the new coordinates (a_1', a_2') of the point A are the components of a vector of length $|\vec{A}|$ and direction angle $\theta_A - \alpha$. Hence, particularly if your calculator sports those special keys for converting from rectangular to polar coordinates and back again, such a change of coordinates by rotation is easily carried out by (i) calculating $|\vec{A}|$ and θ_A from a_1 and a_2, and then (ii) calculating a_1' and a_2' from $|\vec{A'}| = |\vec{A}|$ and $\theta_{A'} = \theta_A - \alpha$.

Problem 1. 6. For each of the following points, calculate their new coordinates after a counterclockwise rotation of the coordinate system by 45°:

(a) $(1, 0)$; (b) $(2, -3)$; (c) $(-10, 1)$; (d) $(-1, -3)$.

2. The car racing game.

This game is a good way to become comfortable with vectors.

The game is played on graph paper on which someone has laid out a racing course, with start and finish line, for example as in Figures 2. 1 and 2. 2. A car's position is given by its position vector, $\vec{P}$, with respect to a coordinate system someone has indicated on the graph paper. The initial position for each car is chosen somewhere on the starting line. In addition, each car has a velocity vector, $\vec{v}$, giving its current velocity. The initial velocity vector is zero.

This means that one must remember four numbers per car (from one move to the next).

Time is discrete in the game, and a move consists in going from $\vec{P}(t)$ to $\vec{P}(t+1)$ according to the formula

$$\vec{P}(t+1) = \vec{P}(t) + \frac{1}{2}(\vec{v}(t) + \vec{v}(t+1)) . \tag{2.1}$$

The velocity vector at time $t+1$ is obtained by

$$\vec{v}(t+1) = \vec{v}(t) + \vec{a} , \tag{2.2}$$

with the acceleration vector $\vec{a}$ chosen by the player for each move (presumably based on the current situation). In effect, each move consists of a unit of time during which each player subjects his or her car to a certain constant acceleration.

The acceleration vector $\vec{a}$ is to be specified by giving its length and its direction angle, i. e. , one specifies the two numbers r and θ, and $\vec{a}$ is then constructed as

$$\vec{a} = r(\cos\theta\ \vec{i} + \sin\theta\ \vec{j}).$$

Figure 2. 1

t	$\vec{P}$	$\vec{v}$	$\vec{a}$
0	(450, 0)	(0, 0)	(10, 90°)
1	(450, 5)	(0, 10)	(10, 90°)
2	(450, 20)	(0, 20)	(10, 90°)
3	(450, 45)	(0, 30)	(10, 90°)
4	(445, 75)	(-10, 30)	(10, 180°)

2. The car racing game

Figure 2. 2

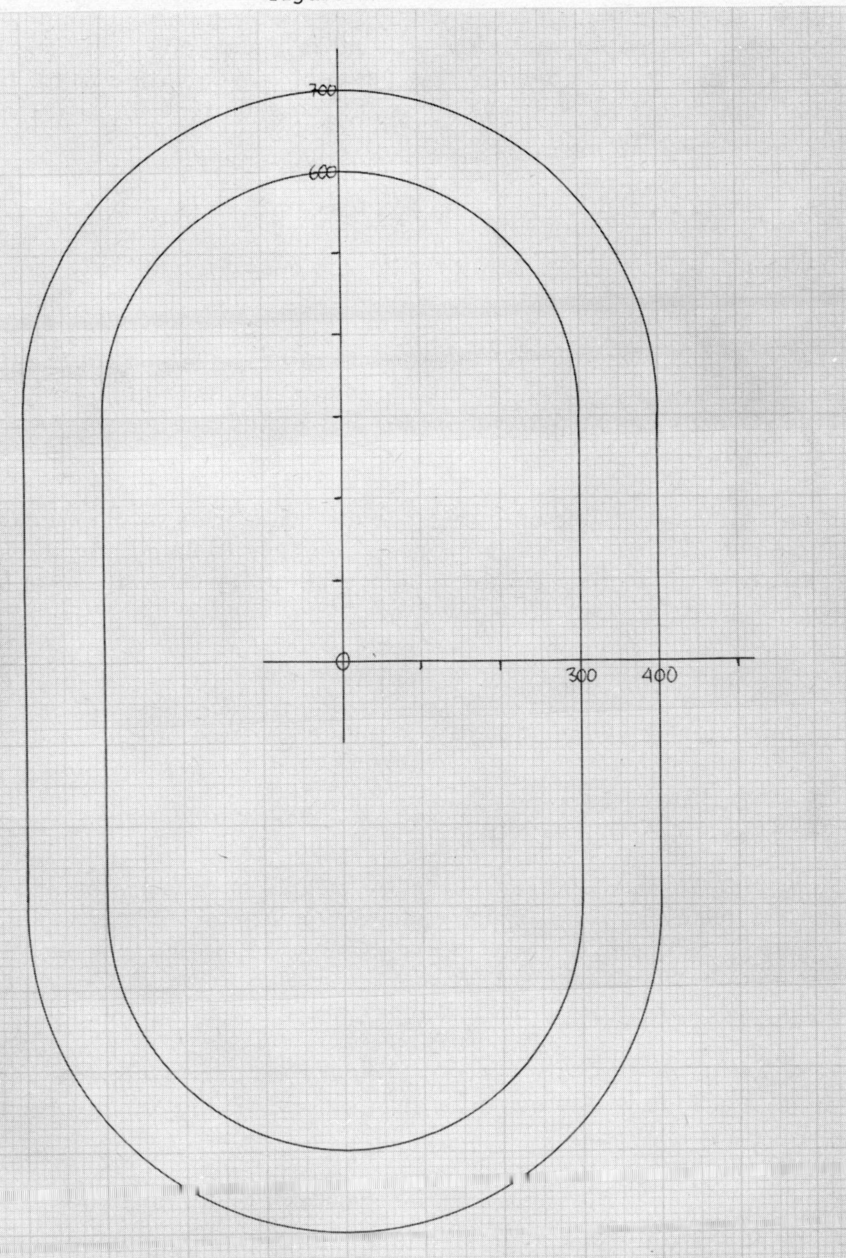

RACE COURSE BOUNDED ON TOP BY SEMICIRCLES WITH CENTER (0,300) AND RADII OF 300 FEET AND 400 FEET; ON BOTTOM SIMILARLY, BUT WITH CENTER AT (0,-300).

START ON X-AXIS WITH 300 < X < 400. GO TWICE AROUND.

We expect the player here to specify θ in DEGREEs.

The game is more interesting if a bound is placed on r. In an automobile, one cannot usually get r much over 10 feet per $\sec^2$, even with a motor that is powerful enough to "burn rubber." So let us agree that in each time interval the player can choose r subject to

(2. 3) $$0 \leq r \leq 10 ,$$

and can choose θ at will (this allows the options of using brakes and skidding around corners).

The convenient way to play the game is to have the current coordinates of $\vec{P}$ and $\vec{v}$ stored at four memory registers, and to have a program which accepts the acceleration $\vec{a} = (r, \theta)$ as input, and then calculates and stores the new values of $\vec{P}$ and $\vec{v}$ according to (2. 1) and (2. 2).

If any value of $\vec{P}$ lies on or outside the boundaries of the race course, the car is deemed to have crashed. If, as proposed, the game is being played on graph paper, one decides this by plotting $\vec{P}$. For the race course of Fig. 2. 1, it is not necessary to use graph paper. Just check each time whether

$$400 < |\vec{P}| < 450 .$$

Actually, for the race course of Fig. 2. 2 one can write a program that will tell if $\vec{P}$ is inside the boundaries or not.

*<u>Problem 2. 1.</u> Write such a program.

<u>Problem 2. 2.</u> Get together with some friends or classmates and try the following variant of the game. Assign somehow starting points on the starting line, or let each player choose. Each player chooses his or her accelerations as the game progresses. Ignore the possibility of two cars crashing into each other. The players take turns advancing their cars one move. See who gets to the finish first without crashing a car.

<u>Remark.</u> If you keep $|\vec{v}|$ small enough, you can insure not crashing, but somebody with a larger $|\vec{v}|$ may get there first. If everybody crashes the first time you try it, then make a fresh start.

<u>Problem 2. 3.</u> Play the game against yourself by trying to get to the finish line in as few moves as possible.

<u>Remark.</u> If you keep a record of the starting point and the accelerations, you can always see what you did by merely running through the game again.

<u>Problem 2. 4.</u> How should you choose the direction of the acceleration vector $\vec{a}$ so as to change the direction of your car as much as possible?

Problem 2. 5. Lay out a different race course, possibly a figure 8, and try the game on it.

Remark. It would be good strategy to design the course so that you can write a program for your calculator to tell you if your car is inside or outside the course.

One can introduce various fine points. We have said that a car did not crash if it was inside the boundary at the end of each move. But it may well have sideswiped the boundary during the move. Or, if we really wish to know if two cars, on the course together, might crash, we have to know where they go during their moves.

Problem 2. 6. Suppose a body is moving in the plane subject to the constant acceleration $\vec{a}$, and let $\vec{P}(s)$ and $\vec{v}(s)$ be its position vector and its velocity vector, respectively, at time s.

(a) Prove that then

$$\vec{P}(t+s) = \vec{P}(t) + s\vec{v}(t) + (s^2/2)\vec{a}, \quad \text{for all } s \text{ and } t. \tag{2.4}$$

(b) Derive (2. 1) and (2. 2) from (2. 4).

(c) Under what circumstances could (2. 4) be used in the game to settle an argument?

3. The angle between two vectors in the plane.

If you need only the direction angle θ_A and not the length $|\vec{A}|$ of the plane vector $\vec{A} = a_1\vec{i} + a_2\vec{j}$, and your calculator has a $\tan^{-1}$ key, then you can calculate θ_A from the equation $\tan\theta_A = a_2/a_1$ which gives

$$\theta_A = \tan^{-1}(a_2/a_1) . \tag{3.1}$$

Of course, you will have to pay attention to the fact that the $\tan^{-1}$ key implements the inverse function for $\tan\theta$ with θ restricted to the interval $-90^\circ < \theta < 90^\circ$; see Fig.3.1. Thus, with $\tan^{-1}$ standing for the function implemented on your calculator,

$$\tan^{-1}(\tan\theta_A) = \theta_A + \pi, \quad \text{if} \quad -\pi < \theta_A < -\frac{\pi}{2}$$

$$\tan^{-1}(\tan\theta_A) = \theta_A - \pi, \quad \text{if} \quad \frac{\pi}{2} < \theta_A < \pi .$$

This means that

$$\theta_A = \begin{cases} \tan^{-1}(a_2/a_1) & \text{if } a_1 > 0 \\ \tan^{-1}(a_2/a_1) - 180^\circ, & \text{if } a_1 < 0 \text{ and } a_2 \geq 0 \\ \tan^{-1}(a_2/a_1) + 180^\circ, & \text{if } a_1 < 0 \text{ and } a_2 < 0 \end{cases} \tag{3.2}$$

Figure 3. 1

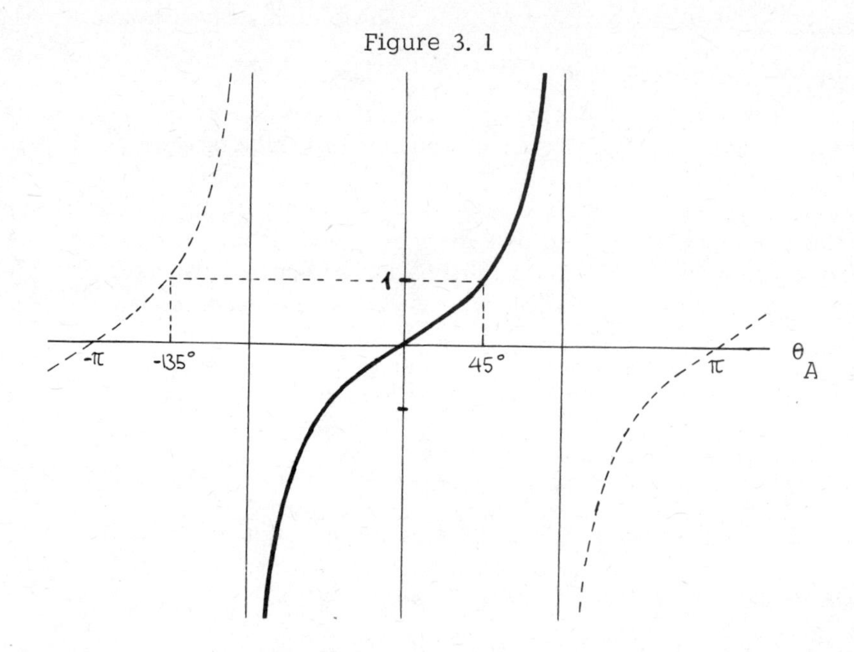

Problem 3. 1. Determine the direction angle for each of the vectors in Prob. 1. 1, but using (3. 2).

Note. If your calculator has a special key for converting from rectangular to polar coordinates, then using (3. 2) for finding θ_A is a waste of your time.

Problem 3. 2. Use (1. 3) or (3. 2) to determine φ, the angle of inclination of the straight line through the points A(-1, -2) and B(2, -1). This is the angle which the straight line makes with the positive x-axis. (Note that $\varphi = \theta_C$, with $\vec{C}$ the vector from A to B.)

Problem 3. 3. You may be used to determining φ, the angle of inclination of a straight line, from the fact that $\tan \varphi = m =$ slope of the straight line. If you calculate φ as $\tan^{-1} m$, do you have to worry about the range of $\tan^{-1}$ as implemented on your calculator?

For any two plane vectors

$$\vec{A} = a_1\vec{i} + a_2\vec{j} \quad \text{and} \quad \vec{B} = b_1\vec{i} + b_2\vec{j},$$

for example as in Fig. 3. 2, we denote by $\theta_{A,B}$ the angle from $\vec{A}$ to $\vec{B}$ measured counter clockwise. This means that

$$\theta_{A,B} = \theta_B - \theta_A, \tag{3. 3}$$

with θ_A and θ_B the direction angles for $\vec{A}$ and $\vec{B}$ respectively, as defined in Sect. 1, provided we do not distinguish between angles that differ by 2π.

3. The angle between two vectors in the plane

Figure 3. 2

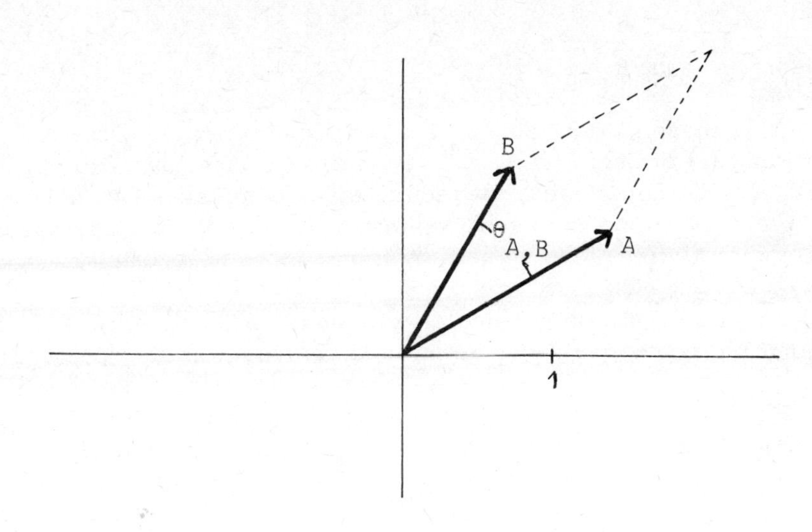

Problem 3. 4. Calculate $\theta_{A, B}$ for the vector pairs:

(a) $\vec{A} = 3\vec{i} + \vec{j}$, $\vec{B} = 4\vec{i} + 5\vec{j}$;

(b) $\vec{A} = -4\vec{i} - \vec{j}$, $\vec{B} = 4\vec{i} - 5\vec{j}$.

Problem 3. 5. Use (3. 3) to calculate the interior angles of the triangle whose vertices are A(1, 1), B(3, -1), C(5, 2). Verify that the sum of the angles is 180° (to within calculator accuracy).

Remark. This is Prob. 17 for Sect. 8 - 2 of T - F.

Hint. To calculate $\measuredangle$ ABC, consider the vector from B to A and the vector from B to C; etc. You should be able to do the assignment by using (1. 3) or (3. 2) just three times.

Problem 3. 6. Calculate the angle between the straight line through the points A(1, 2) and B(2, -1), and the straight line through the points C(1, -4) and D(-1, 2), using (3. 3).

The angle $\theta_{A, B}$ from $\vec{A}$ to $\vec{B}$ figures in two useful equations:

(3. 4) $$|\vec{A}|\,|\vec{B}| \sin \theta_{A, B} = a_1 b_2 - a_2 b_1$$

(3. 5) $$|\vec{A}|\,|\vec{B}| \cos \theta_{A, B} = a_1 b_1 + a_2 b_2$$

Here, (3. 4) gives the (signed) area of the parallelogram spanned by the vectors $\vec{A}$ and $\vec{B}$, and (3. 5) gives the scalar product, or dot product

$$\vec{A} \cdot \vec{B} = a_1b_1 + a_2b_2$$

of the two vectors $\vec{A}$ and $\vec{B}$.

T-F discuss these matters only in the context of vectors in space. Equation (3.4) follows from (1) and (8) of Sect. 11-7 of T-F, while (3. 5) follows from (1) and (6) in Sect. 11-6 of T-F. (In both cases, we reduce the 3-dimensional vectors of T-F to vectors in the plane by taking the z-components to be zero.) But it is possible to verify these equalities directly for plane vectors by high school trigonometry.

By (3. 5), $\vec{A}$ and $\vec{B}$ are perpendicular to each other if and only if $\vec{A} \cdot \vec{B} = 0$. The words "orthogonal" and "normal" are often used as synonyms for "perpendicular."

<u>Problem 3. 7.</u> For the vectors $\vec{A}$ and $\vec{B}$ of Prob. 3. 4. (a), use $\theta_{A,B}$ as calculated in that problem to verify equations (3. 4) and (3. 5) directly to within calculator accuracy. Also, verify that (3. 4) gives the area of the parallelogram spanned by $\vec{A}$ and $\vec{B}$.

<u>Problem 3. 8.</u> Calculate the area of the triangle of Prob. 3. 5, using (3. 4).

<u>Hint.</u> The area of that triangle is half of the area of the parallelogram spanned by the vector from B to A and the vector from B to C.

Equation (3. 5) provides us with an alternative to (3. 3) if we wish to calculate $\theta_{A,B}$. By (3. 5),

$$(3.6) \qquad \theta_{A,B} = \cos^{-1} \frac{\vec{A} \cdot \vec{B}}{|\vec{A}|\,|\vec{B}|} = \cos^{-1} \frac{a_1b_1 + a_2b_2}{\sqrt{(a_1^2 + a_2^2)(b_1^2 + b_2^2)}} .$$

But you would have to pay close attention here, since you may have to change the sign of the result of using the $\cos^{-1}$ key on your calculator. Also, assuming you have already calculated $|\vec{A}|$ and $|\vec{B}|$, calculation of $\theta_{A,B}$ by (3. 6) requires five arithmetic operations and one stroke of the $\cos^{-1}$ key, whereas calculation of $\theta_{A,B}$ by (3. 5) under the same circumstances requires only three arithmetic operations and two strokes of the $\cos^{-1}$ key. Of course, if your calculator has that special key for converting from rectangular to polar coordinates, it is even more convenient to use (3. 3) rather than (3. 6).

The scalar product

$$\vec{A} \cdot \vec{B} = a_1b_1 + a_2b_2$$

is particularly handy for calculating <u>projections</u>, i. e., the decomposition of the vector $\vec{B}$ into the sum of two perpendicular vectors, one of which is parallel to $\vec{A}$. Such a

3. The angle between two vectors in the plane

Figure 3. 3

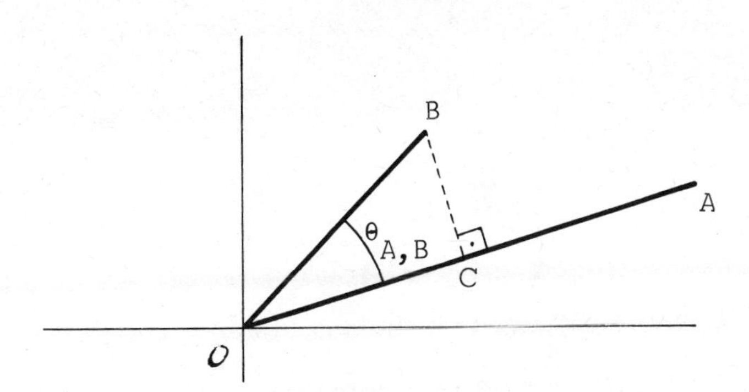

decomposition is obtained as in Fig. 3. 3, where C is the point on the straight line through the origin and A where the perpendicular from B hits. Let $\vec{A}, \vec{B}, \vec{C}$ be the vectors from the origin to the points A, B, C, respectively. Then, obviously,

$$\vec{B} = \vec{C} + (\vec{B} - \vec{C})$$

while $\vec{C}$ and $(\vec{B} - \vec{C})$ are at right angles, and $\vec{C}$ is a scalar multiple of $\vec{A}$. The vector $\vec{C}$ is called the (perpendicular) <u>projection</u> of $\vec{B}$ <u>onto</u> (the direction of) $\vec{A}$, which is written

$$\vec{C} = \operatorname{proj}_{\vec{A}} \vec{B} .$$

We see from Fig. 3. 3 that

$$\vec{C} = (|\vec{B}| \cos \theta_{A, B}) \operatorname{dir} (\vec{A}) .$$

Since $|\vec{A}|\,|\vec{B}| \cos \theta_{A, B} = \vec{A} \cdot \vec{B}$, by (3. 5), this says that

$$\vec{C} = \operatorname{proj}_{\vec{A}} \vec{B} = \frac{\vec{A} \cdot \vec{B}}{|\vec{A}|} \operatorname{dir} (\vec{A}) = \frac{\vec{A} \cdot \vec{B}}{\vec{A} \cdot \vec{A}} \vec{A} . \tag{3. 7}$$

Another way to derive (3. 7) is as follows. Since $\vec{C}$ is in the direction of $\vec{A}$, it must be $\alpha\vec{A}$ for some scalar α. But we wish $\vec{B} - C$ to be perpendicular to $\vec{A}$. That is, we wish $\vec{B} - \alpha\vec{A}$ to be perpendicular to $\vec{A}$. So their scalar product must be zero. That is

$$0 = \vec{A} \cdot (\vec{B} - \alpha\vec{A}) = \vec{A} \cdot B - \alpha(\vec{A} \cdot \vec{A}) .$$

Solving for α gives

$$\alpha = \frac{\vec{A} \cdot \vec{B}}{\vec{A} \cdot \vec{A}} ,$$

as is given in (3. 7).

Here the number $|\vec{B}| \cos \theta_{A, B} = \vec{A} \cdot \vec{B} / |\vec{A}|$ which multiplies the unit vector $\operatorname{dir}(\vec{A}) = \vec{A}/|\vec{A}|$ is called <u>the component of $\vec{B}$ in the direction of $\vec{A}$</u>. Its absolute value is, of course, the length of $\vec{C}$. That is

(3. 8) $$|\vec{C}| = |\operatorname{proj}_{\vec{A}} \vec{B}| = |\vec{A} \cdot \vec{B}| / |\vec{A}| .$$

By construction,

(3. 9) $$\vec{E} = \vec{B} - \vec{C}$$

is perpendicular to $\vec{A}$, and hence to $\vec{C}$. By Fig. 3, 3, its length is

(3. 10) $$|\vec{E}| = |\vec{B}| \, |\sin \theta_{A, B}| ,$$

and this is the distance of the point B from the straight line through the origin and A. This can conveniently be calculated by subtracting $\vec{C}$ from $\vec{B}$, and then taking the length. Or one can appeal to Pythagoras to verify that

(3. 11) $$|\vec{E}|^2 = |\vec{B}|^2 - |\vec{C}|^2 .$$

<u>Problem 3. 9.</u> For $\vec{A}$ and $\vec{B}$ as in Prob. 3.4. (a), calculate $\operatorname{proj}_{\vec{A}} \vec{B}$. Then calculate $\theta_{A, E}$ for the vector $\vec{E} = \vec{B} - \operatorname{proj}_{\vec{A}} \vec{B}$ in order to verify that $\vec{A}$ and $\vec{E}$ are at right angles.

<u>Problem 3. 10.</u> What are the velocity components in the NE direction and the NW direction for an airplane which moves at ground speed of 320 knots in a course of 15° East of North?

<u>Problem 3. 11.</u> For each of the points in Prob. 1. 6, draw the vector from the origin to the point, then calculate its components in the directions of $\vec{i} + \vec{j}$ and of $-\vec{i} + \vec{j}$ and compare with your answers to Prob. 1. 6.

4. The angle between two vectors in space.

We follow T- F and write the typical vector $\vec{A}$ in 3-space as

$$\vec{A} = a_1 \vec{i} + a_2 \vec{j} + a_3 \vec{k} ,$$

with $\vec{i}, \vec{j}$, and $\vec{k}$ the unit vectors in the positive directions of the x- , y- , and z-axis, respectively. You can visualize $\vec{A}$ as the vector from the origin to the point $A(a_1, a_2, a_3)$. Such a vector has <u>length</u>

(4. 1) $$|\vec{A}| = \sqrt{a_1^2 + a_2^2 + a_3^2} ,$$

and <u>direction</u>

4. The angle between two vectors in space

(4. 2) $$\operatorname{dir}(\vec{A}) = \frac{\vec{A}}{|\vec{A}|} = (a_1/|\vec{A}|)\vec{i} + (a_2/|\vec{A}|)\vec{j} + (a_3/|\vec{A}|)\vec{k}.$$

But the direction of $\vec{A}$ is not given by <u>one</u> angle anymore, as in the plane. We must therefore rely on the scalar product for the calculation of the angle between two vectors. If $\vec{B}$ is also a 3-vector,

$$\vec{B} = b_1\vec{i} + b_2\vec{j} + b_3\vec{k},$$

then the <u>scalar</u>, or <u>dot</u>, product of $\vec{A}$ and $\vec{B}$ is given by

(4. 3) $$\vec{A}\cdot\vec{B} = a_1b_1 + a_2b_2 + a_3b_3 .$$

See Sect. 11-6 of T-F, especially Eq. (6).

Some calculators have special statistical keys by which you can save a key stroke or two in calculating the scalar product. However, they tie up a lot of memory registers, so that it is not particularly good to use them.

If you are working with vectors and have a programmable calculator, it is probably a good idea to have available a program to calculate the scalar product. Let us assume that $\vec{A}$ and $\vec{B}$ have been stored; a convenient way is to have a_1, a_2, and a_3 in registers one, two, and three respectively and b_1, b_2, and b_3 in registers four, five, and six respectively. Then the program is so easy to write that we leave the details to the reader. The point is to have the program stored, ready for use.

By (1) of Sect. 11-6 of T-F, we have

(4. 4) $$\vec{A}\cdot\vec{B} = |\vec{A}|\,|\vec{B}|\cos\theta_{A,B}$$

where $\theta_{A,B}$ is the angle between $\vec{A}$ and $\vec{B}$.

Since the vectors $\vec{A}$ and $\vec{B}$ are now in 3-space, we have no notion anymore of clockwise or counterclockwise rotation. For this reason, we do not distinguish between $\theta_{A,B}$ and $\theta_{B,A}$ as we did in the plane, but take for $\theta_{A,B}$ the angle between 0° and 180° formed by $\vec{A}$ and $\vec{B}$. Consequently, from (4. 4),

(4. 5) $$\theta_{A,B} = \cos^{-1}\frac{\vec{A}\cdot\vec{B}}{|\vec{A}|\,|\vec{B}|}$$

with $\cos^{-1}$ as implemented on your calculator (see the discussion in Sect. 1).

<u>Problem 4. 1.</u> Find $\theta_{A,B}$ for the vectors

(4. 6) $$\vec{A} = 3\vec{i} + \vec{j} - 2\vec{k}, \qquad \vec{B} = 4\vec{i} + 5\vec{j} + 3\vec{k}.$$

Problem 4. 2. Interpret the vectors $\vec{A}$ and $\vec{B}$ in Prob. 3. 4 as 3-dimensional vectors with a zero z-component and calculate $\theta_{A, B}$ for them by (4. 5). Compare with your answers for Prob. 3. 4.

The perpendicular projection of $\vec{B}$ onto $\vec{A}$ is given as in the plane by

(4. 7)
$$\vec{C} = \text{proj}_{\vec{A}} \vec{B} = \frac{\vec{A} \cdot \vec{B}}{\vec{A} \cdot \vec{A}} \vec{A} .$$

The derivation of this formula in Sect. 3 only made reference to the fact that the dot product satisfies (4. 4), hence is valid for 3-vectors as well. In particular, the vector

(4. 8)
$$\vec{E} = \vec{B} - \text{proj}_{\vec{A}} \vec{B} = \vec{B} - \vec{C}$$

is a vector in the plane spanned by $\vec{A}$ and $\vec{B}$ and perpendicular to $\vec{A}$, and Eqs. (3. 7)-(3. 11) are valid for 3-dimensional vectors as well.

Problem 4. 3. For the vectors (4. 6), find $\text{proj}_{\vec{A}}\vec{B}$, then calculate $\theta_{A, E}$, with $\vec{E}$ as in (4. 8), to show that $\vec{E}$ is at right angles to $\vec{A}$.

Problem 4. 4. For the vectors (4. 6), find the components of the vector

$$\vec{F} = \vec{A} - \text{proj}_{\vec{B}} \vec{A} .$$

Verify that it is perpendicular to $\vec{B}$ by showing that $\vec{B} \cdot \vec{F} = 0$.

Remark. With your program, mentioned earlier, for calculating the scalar product of two vectors loaded and ready to go, you can calculate $\text{proj}_{\vec{A}}\vec{B}$ as follows.

Put the components of $\vec{A}$ into registers one, two, and three, calculating $\vec{A} \cdot \vec{A} = a_1^2 + a_2^2 + a_3^2$ as you go. This would mean the key strokes

RPN
|
RPN
a_1, [STO 1], [x^2], a_2, [STO 2], [x^2], [+], a_3, [STO 3], [x^2], [+].

AE
|
AE
a_1, [STO 1], [x^2], [+], a_2, [STO 2], [x^2], [+], a_3, [STO 3], [x^2], [=].

Then store $\vec{A} \cdot \vec{A}$, in register zero, for example.

Next, put the components of $\vec{B}$ into registers four, five, and six, use your program to calculate $\vec{A} \cdot \vec{B}$, and then divide this by $\vec{A} \cdot \vec{A}$ (stored in register zero), to get

$$\vec{A} \cdot \vec{B} / \vec{A} \cdot \vec{A}$$

into the display. After this, the key strokes

[STO × 1], [STO × 2], [STO × 3]

will bring the components of $\text{proj}_{\vec{A}}\vec{B}$ into registers one, two, and three, according to (4. 7).

Now, if you also wish the vector $\vec{E} = \vec{B} - \text{proj}_{\vec{A}}\vec{B}$, carry out the key strokes

[RCL 1], [STO - 4], [RCL 2], [STO - 5], [RCL 3], [STO- 6]

which gets the components of $\vec{E}$ into registers four, five and six. If you only want $|\vec{E}|$, calculate $\vec{B}\cdot\vec{B} = |\vec{B}|^2$ as you input the components of $\vec{B}$, and then use (3. 11).

You may wish to combine all these steps into <u>one</u> program, for decomposing $\vec{B}$ into $\text{proj}_{\vec{A}}\vec{B}$ and $\vec{E}$, in which case you may as well load both $\vec{A}$ and $\vec{B}$ initially into registers $1, 2, \ldots, 6$, and only then begin the calculations.

<u>Problem 4. 5.</u> Same as Prob. 4. 3, but with

(a) $\vec{A} = 10\vec{i} + 11\vec{j} - 2\vec{k}$ (b) $\vec{A} = 2\vec{i} + 10\vec{j} - 11\vec{k}$

$\vec{B} = \vec{i} + 3\vec{j} + 4\vec{k}$ $\vec{B} = 2\vec{i} + 2\vec{j} + \vec{k}$.

<u>Remark</u>. This is Prob. 2 and Prob. 5, respectively, for Sect. 11- 6 of T- F.

Consider the line L_1 that passes through

(4. 9) (a_1, a_2, a_3) and (b_1, b_2, b_3).

and the line L_2 that passes through

(4. 10) (c_1, c_2, c_3) and (d_1, d_2, d_3) .

Does it make sense to talk about the angle between these two lines? We had a similar situation in Prob. 3. 6. But there all four points were in the xy -plane. So the lines were bound to meet somewhere (unless they coincided or were parallel, in which case one would say that the angle between them is zero), and where they meet one has an angle between them. But in 3 dimensions, two lines can miss each other completely without being anywhere near parallel. However, we can proceed as in Prob. 3. 6. Consider the four points in (4. 9) and (4. 10) as determining vectors $\vec{A}, \vec{B}, \vec{C}$, and $\vec{D}$. Then $\vec{B} - \vec{A}$ is parallel to L_1 and $\vec{D} - \vec{C}$ is parallel to L_2. As these latter two vectors proceed from the origin, we can get the angle $\theta_{B-A,\,D-C}$ between them by (4. 5).

<u>Problem 4. 6.</u> Find the angle between the line L_1 determined by the two points $A(1, 0, -1)$, $B(-1, 1, 0)$ and the line L_2 determined by the points $C(3, 1, -1)$, $D(4, 5, -2)$.

<u>Remark</u>. The next three problems are Problems 3, 6, and 7 at the end of Sect. 11-6 of T- F.

If you are using some text other than T- F, it would be prefectly all right to work similar problems from your own text. However, take note of the remark which follows Prob. 4. 9.

Problem 4. 7. Find the interior angles of the triangle ABC whose vertices are the points $A(-1, 0, 2)$, $B(2, 1, -1)$, and $C(1, -2, 2)$. Verify that the sum of the angles is 180° (to within calculator accuracy).

Problem 4. 8. Find the angle between the diagonal of a cube and one of its edges.

Problem 4. 9. Find the angle between the diagonal of a cube and a diagonal of one of its faces.

Remark. Explain why it is geometrically obvious that the angles in Problems 4. 8 and 4. 9 add up to exactly 90°. So you should not have used the calculator to do Prob. 4. 9. Simply subtract the answer for Prob. 4. 8 from 90°. As we keep saying, stay alert. NEVER work a problem on a calculator if there is some trivial way that you can just write down the answer.

5. The vector product.

The vector product $\vec{A} \times \vec{B}$ of two vectors in space is again a vector in space, defined by

(5. 1) $$\vec{A} \times \vec{B} = \vec{i}(a_2 b_3 - a_3 b_2) - \vec{j}(a_1 b_3 - a_3 b_1) + \vec{k}(a_1 b_2 - a_2 b_1);$$

see Eq. (8) in Sect. 11-7 of T-F. This definition insures that the scalar product of $\vec{A} \times \vec{B}$ with both the vector $\vec{A}$ and the vector $\vec{B}$ is zero, as you can verify directly with a little algebra.

If you already know determinants, then you will find it easier to remember (5. 1) in the form

(5. 2) $$\vec{A} \times \vec{B} = \det \begin{bmatrix} \vec{i} & \vec{j} & \vec{k} \\ a_1 & a_2 & a_3 \\ b_1 & b_2 & b_3 \end{bmatrix} ;$$

see Eq. (9) in Sect. 11-7 of T-F. For here, the components of $\vec{A}$ and $\vec{B}$ appear in a very simple and orderly manner. You will then also realize at once that the scalar product of any vector $\vec{C}$ with $\vec{A} \times \vec{B}$ can be calculated as

(5. 3) $$\vec{C} \cdot (\vec{A} \times \vec{B}) = \det \begin{bmatrix} c_1 & c_2 & c_3 \\ a_1 & a_2 & a_3 \\ b_1 & b_2 & b_3 \end{bmatrix}$$

(see Eq. (3) in Sect. 11-9 of T-F). This makes it explicit that

(5. 4) $$\vec{A} \cdot (\vec{A} \times \vec{B}) = \vec{B} \cdot (\vec{A} \times \vec{B}) = 0$$

since you will then also know that any determinant with two rows the same is automatically zero.

On the other hand, if you do not know determinants, then you will just have to struggle along with Eq. (5. 1) and hope to remember it somehow. You could, of course, take the occasion and learn now how to evaluate a determinant. That knowledge will come in handy anyway, and then you would have to remember only (5. 2).

As an aid in such a worthwhile endeavor, here is a description of how to evaluate a determinant by expansion with respect to the top row. For example, for a 2×2 determinant

$$\det\begin{bmatrix} a_1 & a_2 \\ b_1 & b_2 \end{bmatrix} = a_1 \det\begin{bmatrix} a_1 & a_2 \\ b_1 & b_2 \end{bmatrix} - a_2 \det\begin{bmatrix} a_1 & a_2 \\ b_1 & b_2 \end{bmatrix}$$

$$= a_1 \cdot b_2 \qquad - a_2 \cdot b_1 .$$

For a 3×3 determinant, there are three terms,

$$\det\begin{bmatrix} c_1 & c_2 & c_3 \\ a_1 & a_2 & a_3 \\ b_1 & b_2 & b_3 \end{bmatrix} = c_1 \det\begin{bmatrix} c_1 & c_2 & c_3 \\ a_1 & a_2 & a_3 \\ b_1 & b_2 & b_3 \end{bmatrix} - c_2 \det\begin{bmatrix} c_1 & c_2 & c_3 \\ a_1 & a_2 & a_3 \\ b_1 & b_2 & b_3 \end{bmatrix} + c_3 \det\begin{bmatrix} c_1 & c_2 & c_3 \\ a_1 & a_2 & a_3 \\ b_1 & b_2 & b_3 \end{bmatrix}$$

$$= c_1 \det\begin{bmatrix} a_2 & a_3 \\ b_2 & b_3 \end{bmatrix} - c_2 \det\begin{bmatrix} a_1 & a_3 \\ b_1 & b_3 \end{bmatrix} + c_3 \det\begin{bmatrix} a_1 & a_2 \\ b_1 & b_2 \end{bmatrix}$$

$$= c_1 (a_2 b_3 - a_3 b_2) - c_2 (a_1 b_3 - a_3 b_1) + c_3 (a_1 b_2 - a_2 b_1) .$$

In words: Multiply each entry in the top row by the determinant left over after you have struck out the row and the column of that entry, then change the sign of every other one of these products, and then sum.

The vector (or cross) product $\vec{A} \times \vec{B}$ is used in calculus and engineering as a handy device for constructing a vector which is, by (5. 4), perpendicular to the two vectors $\vec{A}$ and $\vec{B}$, whatever those vectors might be. The vector product is also useful in area calculations because

(5. 5) $\qquad$ area of parallelogram spanned by $\vec{A}$ and $\vec{B}$ $= |\vec{A} \times \vec{B}|$.

See Sect. 11 - 7 of T - F.

Thus, consider the vectors of (4. 6), namely

(5. 6) $$\vec{A} = 3\vec{i} + \vec{j} - 2\vec{k}$$

(5. 7) $$\vec{B} = 4\vec{i} + 5\vec{j} + 3\vec{k}\ .$$

By (5. 1), we have

$$\vec{A} \times \vec{B} = \vec{i}((1)(3) - (-2)(5))$$
$$- \vec{j}((3)(3) - (-2)(4))$$
$$+ \vec{k}((3)(5) - (1)(4))\ .$$

So

(5. 8) $$\vec{A} \times \vec{B} = 13\vec{i} - 17\vec{j} + 11\vec{k}\ .$$

By (5. 5), the area of the parallelogram spanned by $\vec{A}$ and $\vec{B}$ is $|\vec{A} \times \vec{B}| = \sqrt{579}$. But this is twice the area of the triangle of which $\vec{A}$ and $\vec{B}$ are two sides. So the area of said triangle is $\sqrt{579}/2 \cong 120.31$.

We can verify that

(5. 9) $$(\vec{A} \times \vec{B}) \times \vec{B} = -106\vec{i} + 5\vec{j} + 133\vec{k}\ .$$

Problem 5. 1. Explain geometrically why the vector $(\vec{A} \times \vec{B}) \times \vec{B}$ should be parallel to the vector $\vec{F}$ obtained in Prob. 4. 4. From the answer for Prob. 4. 4, verify that the two vectors are indeed parallel.

If you have a programmable calculator, it would be a good idea to construct and keep a program for calculating $\vec{A} \times \vec{B}$. As with the program for calculating the scalar product $\vec{A} \cdot \vec{B}$, and the program for calculating the projection of $\vec{B}$ onto $\vec{A}$, start with having the components of $\vec{A}$ in registers one, two, three, and those of $\vec{B}$ in registers four, five, six. Then calculate the first component of $\vec{A} \times \vec{B}$ and store it temporarily in register 0, calculate the second component of $\vec{A} \times \vec{B}$ and store it temporarily in register seven. Finally, calculate the third component of $\vec{A} \times \vec{B}$ and store it in register three, and now move the first component from register zero to register one, and the second component from register seven to register two. Here is a program.

RPN

[RCL 2], [RCL 6], [×], [RCL 3], [RCL 5], [×], [−], [STO 0],
[RCL 3], [RCL 4], [×], [RCL 1], [RCL 6], [×], [−],
[RCL 1], [RCL 5], [×], [RCL 2], [RCL 4], [×], [−], [STO 3],
[R↓], [STO 2], [RCL 0], [STO 1].

This takes 27 program steps.

RPN

AE

[RCL 2], [×], [RCL 6], [−], [(], [RCL 3], [×], [RCL 5], [=], [STO 0],
[RCL 3], [×], [RCL 4], [−], [(], [RCL 1], [×], [RCL 6], [=], [STO 7],

AE

AE

[RCL 1], [×], [RCL 5], [−], [(], [RCL 2], [×], [RCL 4], [=], [STO 3],
[RCL 0], [STO 1], [RCL 7], [STO 2].

This takes 34 program steps. Of these, the three parenthesis key strokes may be omitted on a hierarchical calculator such as the TI-57.

AE

This replaces $\vec{A}$ by $\vec{A} \times \vec{B}$ in registers one, two, three, but leaves $\vec{B}$ untouched in registers four, five, six.

After you have stored your program for calculating $\vec{A} \times \vec{B}$, it would not be a bad idea to test it. For this, put the two vectors $\vec{A}$ and $\vec{B}$ of (5. 6) and (5. 7) in place. Form $\vec{A} \times \vec{B}$, and see if it agrees with (5. 8). But now you have $\vec{A} \times \vec{B}$ and $\vec{B}$ right in place to form $(\vec{A} \times \vec{B}) \times \vec{B}$ by your program. Do that, and see if it agrees with (5. 9).

Calculating $\vec{A} \times \vec{B}$ was one of those things that was almost impossible to do correctly before calculators came along. Even with a nonprogrammable calculator it is still very hard. You start by taking the product of the second component of the first vector with the third component of the second vector, and by the time you are half way through you have lost your place, and have blown it. Besides which, half the terms have minus signs in front. Sprinkle a few negative numbers amongst the coefficients of $\vec{A}$ and $\vec{B}$, and you are bound to get a sign wrong some place. But with a carefully checked program on your programmable calculator, there is no trouble whatsoever.

If calculating $\vec{A} \times \vec{B}$ was hard, then calculating $(\vec{A} \times \vec{B}) \times \vec{C}$ was at least twice as hard. So great efforts were made to find easier ways of calculation. Eq. (4) in Sect. 11-9 of T-F gives

(5. 10) $$(\vec{A} \times \vec{B}) \times \vec{C} = (\vec{A} \cdot \vec{C})\vec{B} - (\vec{B} \cdot \vec{C})\vec{A}.$$

The right side of this is not all that easy to calculate, but before calculators it was an improvement over the left side. But with a tested program for calculating the vector product, the left side of (5. 10) is suddenly very easy, while the right side of (5. 10) is still a mess. Besides which, how could one possibly ever remember it ?

So, with a programmable calculator, you can forget about formulas such as (5. 10). Not to mention the dreadful expression that some calculus books get for

$$(A \times B) \times (C \times D).$$

One minor point. The program we suggested for calculating $\vec{A} \times \vec{B}$ kept $\vec{B}$ but washed out $\vec{A}$. Suppose you wish to keep $\vec{A}$. Just interchange $\vec{A}$ and $\vec{B}$. Of course, then you have calculated $\vec{B} \times \vec{A}$. However,

$$\vec{A} \times \vec{B} = -(\vec{B} \times \vec{A}),$$

so that at the end you change the signs of the components of $\vec{B} \times \vec{A}$.

If you wish to keep both $\vec{A}$ and $\vec{B}$ without having to record any coefficients in writing, you will have to get a calculator with more than eight memory registers.

Remark. The next five problems are Problems 1, 2, 3, 4, and 7 at the end of Sect. 11-7 of T-F.

If you are using some text other than T-F, it would be perfectly all right to work similar problems from your own text.

Problem 5.2. Find $\vec{A} \times \vec{B}$ if $\vec{A} = 2\vec{i} - 2\vec{j} - \vec{k}$ and $\vec{B} = \vec{i} + \vec{j} + \vec{k}$.

Problem 5.3. Find a vector $\vec{N}$ perpendicular to the plane determined by the points A(1, -1, 2), B(2, 0, -1), and C(0, 2, 1).

Problem 5.4. Find the area of the triangle ABC of Prob. 5.3.

Problem 5.5. Find the distance between the origin and the plane ABC of Prob. 5.3 by projecting $\overrightarrow{OA}$ onto the perpendicular vector $\vec{N}$ of Prob. 5.3.

Problem 5.6. Using vector methods, find the distance between the line L_1 determined by the two points A(1, 0, -1), B(-1, 1, 0) and the line L_2 determined by the points C(3, 1, -1), D(4, 5, -2). The distance is to be measured along a line perpendicular to both L_1 and L_2.

Hint. Take a vector $\vec{A}$ parallel to L_1 and a vector $\vec{B}$ parallel to L_2. Then $\vec{A} \times \vec{B}$ has the direction of the line perpendicular to both L_1 and L_2. Take $\vec{C}$ a vector connecting a point on L_1 with one on L_2, and find the length of its perpendicular projection onto $\vec{A} \times \vec{B}$.

Remark. The next two problems are Problems 8 and 16 for Sect. 11-8 of T-F. If you are using some text other than T-F, it would be perfectly all right to work similar problems for your own text. The key points to remember are that the equation of a plane in space is of the form

$$ax + by + cz = d$$

with $a\vec{i} + b\vec{j} + c\vec{k}$ a vector perpendicular to the plane, and that the equations of a straight line in space are of the form

$$\frac{x - x_0}{a} = \frac{y - y_0}{b} = \frac{z - z_0}{c}$$

with (x_0, y_0, z_0) a point on the line and $a\vec{i} + b\vec{j} + c\vec{k}$ a vector parallel to that line.

Problem 5.7. Find a plane through the points A(1, 1, -1), B(2, 0, 2), C(0, -2, 1).

Problem 5.8. Show that the line of intersection of the planes

$$x + 2y - 2z = 5 \quad \text{and} \quad 5x - 2y - z = 0$$

is parallel to the line

$$\frac{x+3}{2} = \frac{y}{3} = \frac{z-1}{4} .$$

Find the plane determined by these two lines.

6. Volume of a tetrahedron.

Suppose three vectors $\vec{A}$, $\vec{B}$, and $\vec{C}$ proceed from the origin. According to the second paragraph of Sect. 11-9 of T-F,

$$(\vec{A} \times \vec{B}) \cdot \vec{C} \tag{6.1}$$

is, except perhaps for sign, the volume of a parallelepiped of which $\vec{A}$, $\vec{B}$, and $\vec{C}$ are three concurrent edges. Hence (6.1) is, except perhaps for sign, six times the volume of the tetrahedron whose vertices are the origin, A, B, and C.

To carry out the calculation of (6.1), we store $\vec{A}$ and $\vec{B}$ on the calculator. Then run the program for calculating the vector product. Now $\vec{A} \times \vec{B}$ will be where $\vec{A}$ was. Then put $\vec{C}$ in place of $\vec{B}$ and run the program for calculating the scalar product.

Problem 6.1. Find the volume of the tetrahedron with vertices at (0, 0, 0), (1, -1, 1), (2, 1, -2), and (-1, 2, -1).

Remark. This is Prob. 6 at the end of Sect. 11-9 of T-F.

Problem 6.2. Find the volume of the tetrahedron with vertices at (1, 0, -1), (-1, 1, 0), (3, 1, -1), and (4, 5, -2).

Chapter XII

FUNCTIONS OF SEVERAL VARIABLES

0. Guide for the reader.

You should read Sect. 1 when functions of two variables are discussed in your course, since it provides a quick way to sketch such functions. Sect. 2 deals with tangent planes, and you should read it as soon as you have come across derivatives for a function of several variables.

1. Plotting a function of two variables.

It is very hard to sketch a function of more than one variable. If the function depends on more than two variables, we would have to visualize something involving at least four dimensions, and that we cannot do. But even for a function of just two variables, it isn't easy. A function f of two variables x and y is usually visualized as a surface in 3-space, namely the surface made up of the points

$$(x, y, f(x, y)) .$$

But to model such a surface as a three-dimensional object takes a great deal of effort, particularly if it is to be done accurately. The next best thing is a drawing, that is, a two-dimensional representation of this three-dimensional object. It still takes a craftsman (or, these days, a computer) to make an accurate perspective drawing of such a surface $z = f(x, y)$, so we are looking for something simpler than that.

In industry, one usually uses <u>sections</u>. These are functions of <u>one</u> variable, obtained from f by keeping one of the variables fixed, for example the functions g_j given by

$$g_j(x) = f(x, y_j)$$

or the functions h_i given by

$$h_i(y) = f(x_i, y) .$$

1. Plotting a function of two variables

These are plotted in the usual way, and a sequence of them can be used to get a feeling for the function f.

Sections are used because it is relatively easy to get information about the function out of them. But a more comprehensive view of such a function is to be had from a map for it. By this we mean a topographic map, such as is found in many atlases. This requires you to sketch, in the (x, y)-plane, selected level curves of the function f. To recall, the level curve C_α for the level α consists of all points (x, y) in the plane for which $f(x, y) = \alpha$,

$$C_\alpha = \{(x, y) : f(x, y) = \alpha\} .$$

You would plot C_α restricted to some rectangle (or other region) in the domain of F, and for certain discrete levels $\alpha_1 < \alpha_2 < \ldots < \alpha_N$ For best effect, these levels should be uniformly spaced in α, as they are on a topographic map. For then, the closer the level curves are, the steeper the function is there. The discussion in Sect. 13-2 of T-F is very relevant here.

It is, of course, possible, to do a beautiful job on fine graph paper from finely spaced data. These days, it is also possible to use canned programs on bigger computers to generate and then plot level curves for you. But you can produce easily rough plots yourself, with just a little bit of effort and a calculator.

For a start, evaluate $f(x, y)$ at all the points of a rectangular grid, say at the points (x_i, y_j) with

$$x_i = x_0 + ih, \quad i = 0, \ldots, M$$

$$y_j = y_0 + jk, \quad j = 0, \ldots, N .$$

You might as well have the mesh spacing in x the same as in y,

$$h = k,$$

unless there is a good reason not to. Then, on a piece of graph paper, draw an appropriate piece of the x-axis and of the y-axis and write, for each i and j, the first two or three significant digits of the number $f(x_i, y_j)$ at the point corresponding to (x_i, y_j).

This we have done in Fig. 1.1, for the specific function f given by

$$f(x, y) = x^3/3 + x(y^2 - 1) - y/2 + 1 \tag{1.1}$$

and for

$$x_0, x_1, \ldots, x_M = -0.2, 0, 0.2, \ldots, 1.2$$

$$y_0, y_1, \ldots, y_N = 0, 0.2, \ldots, 1.2 .$$

Figure 1. 1

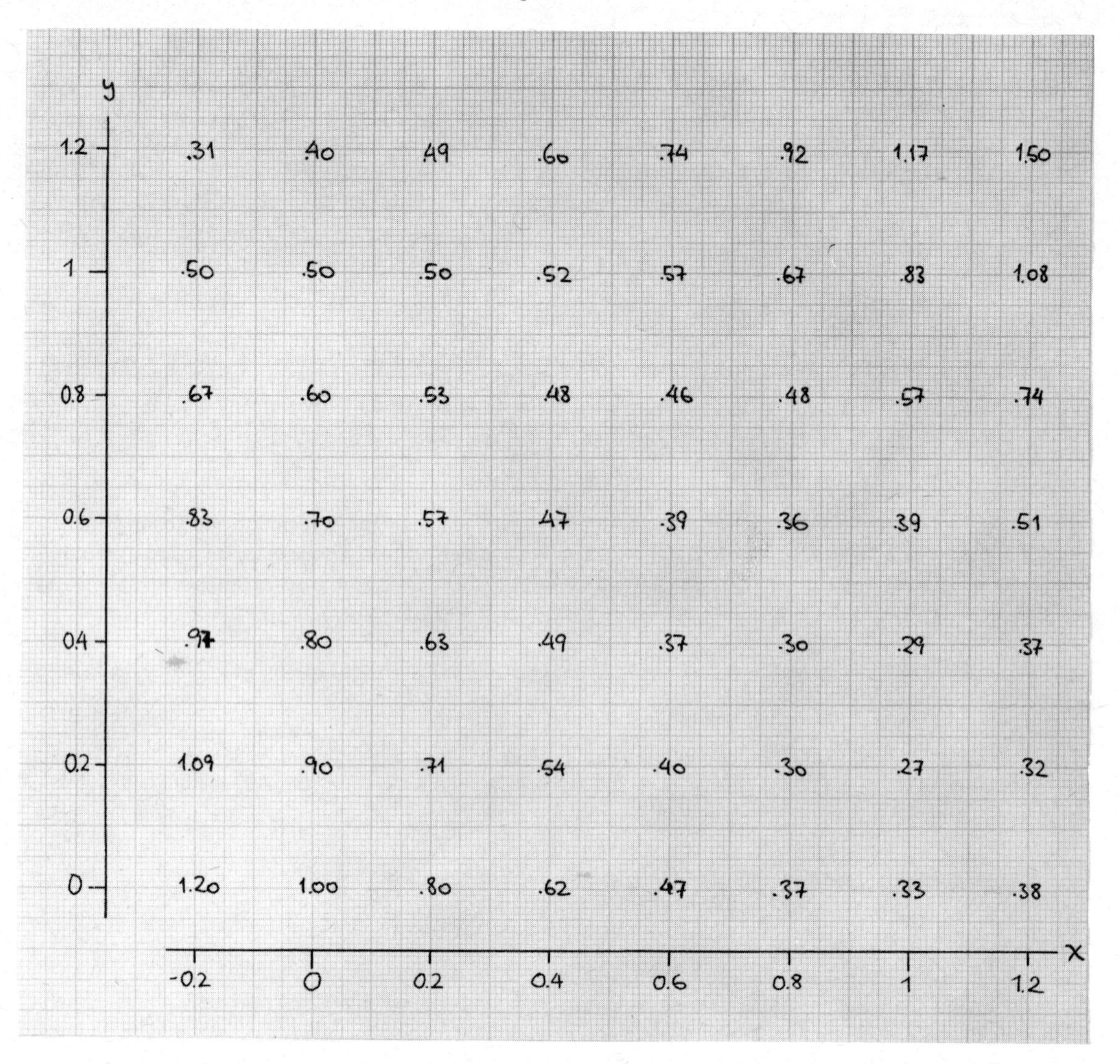

It helps here to have a programmable calculator. One would have a program for evaluating $f(x, y)$, which would use x, y as stored in registers 1, 2 say. The program would begin by adding the step h to register 1. So, every time you push the appropriate program key, x would be incremented by h and $f(x, y)$ evaluated at the current x and y. Supposing y_j stored in register 2, and the number $x_0 - h$ initially in register 1, the repeated execution of the program will give in sequence the numbers $f(x_0, y_j)$, $f(x_1, y_j)$, $f(x_2, y_j), \ldots$, ready to be written down on your graph paper. This fills one row of the grid. Then, to get the next row, increment y_j in register 2 by k to get y_{j+1} there, restore $x_0 - h$ to register 1, and you are ready for the next row of function values.

In this way, you get a table of function values. Some facts about f are already quite evident from such a table. For example, in Fig. 1.1, the smallest value occurs at (1, 0.2), which would indicate that f has a local minimum nearby. The largest value appears at (1.2, 1.2), on the boundary of the plotted region.

But if you desire a more graphical accout of f, you should now sketch in some level curves.

The basic idea for this is as follows. Suppose it is the level α you wish to follow. Then you begin by finding a rectangle

$$R_{i,j}: x_i \le x \le x_{i+1}, \quad y_j \le y \le y_{j+1}$$

for which the maximum of f over the four corners is greater than α while the minimum over the four corners is less than α. If $R_{i,j}$ is such a rectangle, then there must be at least two points on its boundary at which f has the value α exactly, assuming (as we do here) that f is a continuous function. So, find those two points, then connect them with a straight line to give you a sketch of C_α on that rectangle $R_{i,j}$. Of course, this means that the curve C_α goes on into neighboring rectangles. So, repeat the process there. In this way, you pursue the curve C_α from rectangle to rectangle until either you reach again that first rectangle and the curve closes on itself, or else you reach a rectangle on the boundary of your plotting region, and your curve C_α leaves the region there. If you do not like loose ends, start a level curve in a boundary rectangle, if possible.

There is, of course, no point in determining the points where C_α enters and exits a rectangle exactly. If f doesn't change too much from one corner of the rectangle to the other, you can guess the exit points by eye. If you want to be more careful, use inverse interpolation (see Chap. VI), as you would when guessing what x gives a certain y in a table of y's against x's. For example, if $f(x_i, y_j) = z_i > \alpha > z_{i+1} = f(x_{i+1}, y_j)$, then by (4.2) of Chap. VI, an approximation for the x for which $f(x, y_j) = \alpha$ would be given by

$$x - x_i = \frac{\alpha - z_i}{z_{i+1} - z_i}(x_{i+1} - x_i) \; ; \tag{1.2}$$

see Fig. 1.2. Since $x_{i+1} - x_i = h$ does not depend on i, it would be worthwhile to have a little program which would have α and h built in (or stored somewhere) and would calculate $x - x_i$ from input z_i and z_{i+1} by

$$x - x_i = \frac{\alpha - z_i}{z_{i+1} - z_i} h \, . \tag{1.3}$$

If the table spacing in y is the same, i.e., if h = k, then the same program can be used to find

$$y - y_j = \frac{\alpha - z_j}{z_{j+1} - z_j} k \, , \tag{1.4}$$

Figure 1. 2

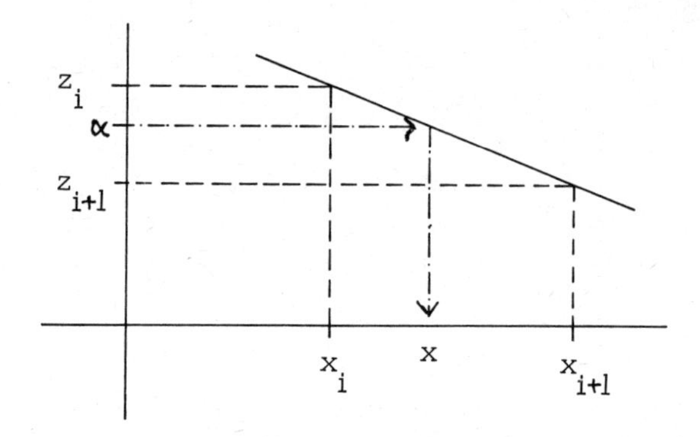

i. e., for finding the point y at which C_α crosses some line $x = x_i$; if $k \neq h$, only a slight change is required in the program.

Let's try all this for our example. We intend to plot the level curve $C_{0.4}$ from the information about f as given in Fig. 1. 1. The boundary rectangle

$$R_{6,2} : 1 \leq x \leq 1.2, \quad 0.4 \leq y \leq 0.6$$

has a maximum corner value 0. 51 and minimum corner value 0. 29, and therefore contains a piece of $C_{0.4}$. One exit point is on its East side, where the two corner values 0. 37 and 0. 51 bracket the level 0. 4. By inverse interpolation, the relevant y-value satisfies

$$y - 0.4 = \frac{0.4 - 0.37}{0.51 - 0.37}(0.2) = 0.04$$

and so you would mark this point on the $x = 1.2$ line, 0. 04 up from the $y = 0.4$ line. Feel free to mark it right on Fig. 1. 1! The other exit is on the North side of the rectangle, since the corner values 0. 39 and 0. 51 there bracket the level 0. 4. The relevant x-value satisfies

$$x - 1 = \frac{0.4 - 0.39}{0.51 - 0.39}(0.2) = 0.02 .$$

So you would mark this point in Fig. 1. 1, on the $y = 0.6$ mesh line, and 0. 02 to the right of the $x = 1$ line. Then, before leaving the rectangle, connect the two exit points with a straight line.

With this, you have marked already one exit of $C_{0.4}$ for the next rectangle, the rectangle

$$R_{6,3} : 1 \leq x \leq 1.2, \quad 0.6 \leq y \leq 0.8 .$$

The other exit is on the West side of this rectangle, between the function values 0.39 and 0.57 written there. The relevant y-value satisfies

$$y - 0.6 = \frac{0.4 - 0.39}{0.57 - 0.39}(0.2) = 0.01 ,$$

and you would mark it, on the mesh line $x = 1$, and 0.01 above the mesh line $y = 0.6$.

This brings you to the rectangle

$$R_{5,3} : 0.8 \le x \le 1, \quad 0.6 \le y \le 0.8$$

and, continuing, into rectangles $R_{4,3}$, $R_{4,2}$ and, finally, into rectangle $R_{4,1}$ where something new happens: The curve $C_{0.4}$ leaves that rectangle at a corner, the SE corner. It enters $R_{5,0}$ and then exits the plotting region at the South side of $R_{5,0}$.

We hope that you carried out all the details of this discussion, marking the curve C_α as you went. If you did, you should now have drawn a level curve $C_{0.4}$ as it appears in Fig. 1.3. There we have drawn C_α in this manner for $\alpha = 0.3, 0.4, \ldots, 1.4$.

Fig. 1.3 gives us a much more detailed feeling of the behavior of f than does Fig. 1.1. We see the level curves $C_{0.5}$, $C_{0.4}$ and $C_{0.3}$ concentrating on a local minimum. We see the level curves more and more densely packed in the upper right corner, indicating a more and more rapid rise of f there. There is a similar but not quite so rapid rise in the lower left corner. In addition, an interesting feature has emerged in the upper left corner, a <u>saddle point</u>, in the rectangle

$$R_{2,4} : 0.2 \le x \le 0.4, \quad 0.8 \le y \le 1 .$$

Of course, we do not maintain that the triangle we ended up drawing there in the rectangle is an accurate description of the level curve $C_{0.5}$ in that rectangle. Since there are more than two exit points for $C_{0.5}$ in that rectangle, we do not know how these points are connected by the curve. But, by connecting them in all possible ways, we have drawn attention to this special spot. For, it can be shown that f has a saddle point in any rectangle in which some level curve has more than two exits.

Finally, the figure strongly supports the view that the maximum value of f on $-0.2 \le x \le 1.2$, $0 \le y \le 1.2$ occurs at (1.2, 1.2), while the minimum value occurs at a local minimum, near (0.95, 0.25). Also, a second critical point, a saddle point, appears to be near (0.3, 0.9).

<u>Problem 1.1.</u> Construct again the level curves for f given by (1.1), but from the rougher mesh

$$x_0, \ldots, x_M = -0.5, 0, 0.5, 1.0, 1.5$$

$$y_0, \ldots, y_N = 0, 0.5, 1, 1.5$$

and compare with Fig. 1.3.

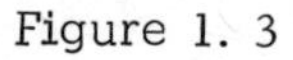
Figure 1. 3

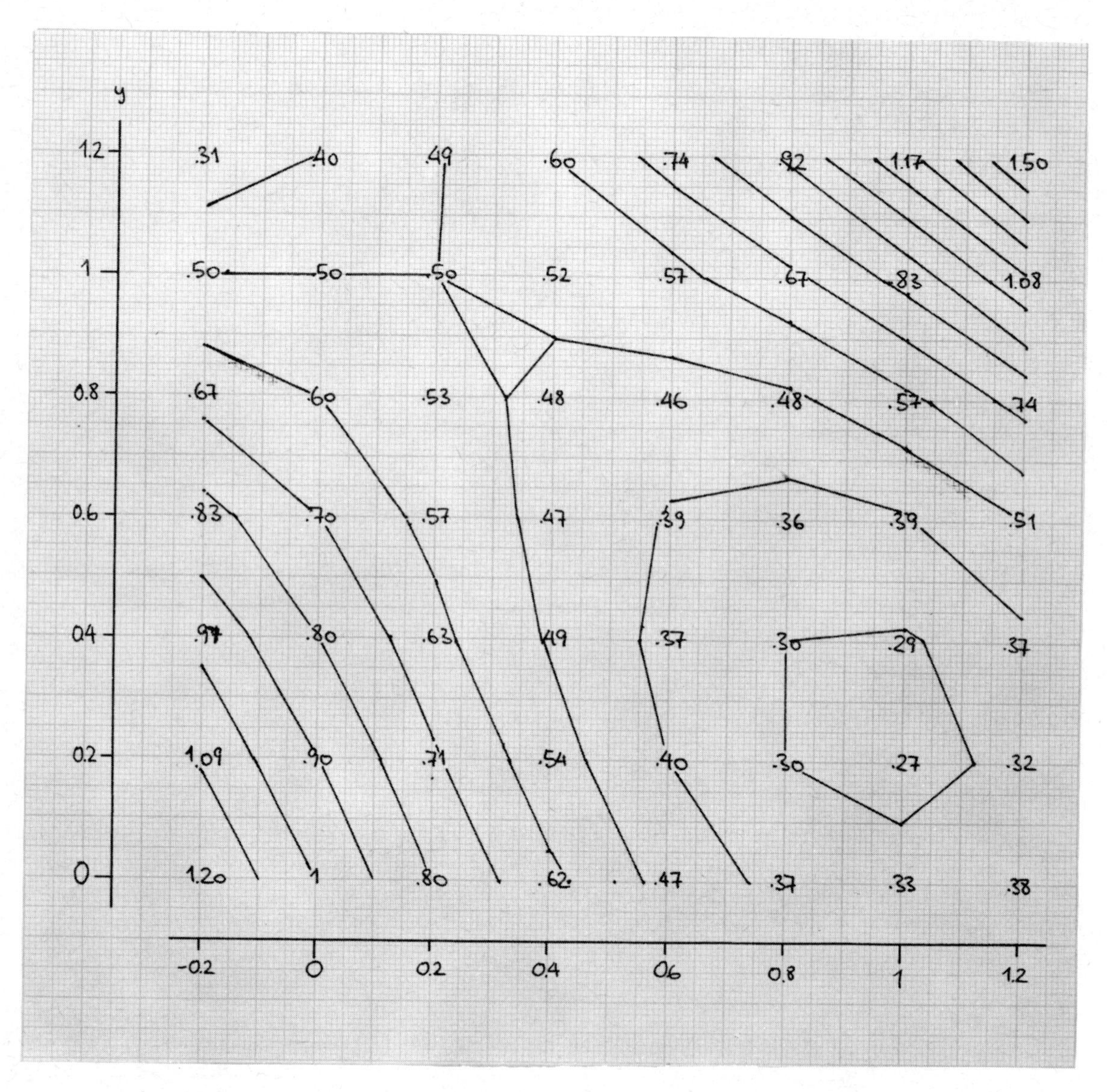

<u>Problem 1. 2.</u> Sketch level curves for f as given by (1. 1), but for the region $-1.2 \le x \le 0$, $-1.2 \le y \le 0$, and determine approximately its minimum and its maximum there. Also, are there any saddle points? Answer these questions first, after you have written down the table of function values, and then again, after you have also plotted the level curves.

<u>Problem 1. 3.</u> Sketch level curves for the function f given by

$$f(x, y) = \frac{\sin (x+y)}{\cos(x-y) + 1.1} \qquad 1 \le x \le 3, \quad -2 \le y \le 0 ,$$

using the mesh step 0.5 in both x and y. From your sketch, determine the approximate location of the maximum and the minimum of f.

Problem 1.4. Sketch level curves for the function f given by

$$f(x,y) = \begin{cases} 0 & \text{if } x = y = 0 \qquad 0 \le x,\ y \le 1, \\ \dfrac{xy}{x^2+y^2}, & \text{otherwise} \end{cases}$$

using a mesh step of 0.2 in both x and y. Then determine analytically the level curves C_0 and $C_{0.5}$ and compare with those you have drawn. Conclude that the technique discussed in this section will not reveal a discontinuity.

Remark. This discontinuity is discussed in Example 2 of Sect. 13-1 of T-F.

2. The tangent plane.

For a function f of several variables $x, y, \ldots, z$, the tangent plane at a point $(x_0, y_0, \ldots, z_0)$ is described by the linear function T given by

$$(2.1) \qquad T(x, y, \ldots, z) = f + f_x \cdot (x - x_0) + f_y \cdot (y - y_0) + \ldots + f_z \cdot (z - z_0)$$

with each of the functions $f, f_x, f_y, \ldots, f_z$ evaluated at the same point $(x_0, y_0, \ldots, z_0)$. Here, we have written f_x for the partial derivative of f with respect to the first variable, f_y for the partial derivative of f with respect to the second variable, etc. See Eq. (32) in Sect. 13-4 of T-F.

For such a function f, the linear function T takes over the role which the tangent line plays for a function of one variable. In particular, our discussion in Chap. VI of the tangent line as a local approximation to f is applicable to the tangent plane T as well.

The function T is characterized by the fact that it agrees with f at $(x_0, y_0, \ldots, z_0)$ in value, and in every directional derivative. As a consequence, its value at some point $(x, y, \ldots, z)$ "near" $(x_0, y_0, \ldots, z_0)$ gives an indication of what the value of f at that point $(x, y, \ldots, z)$ might be. This indication is the more reliable, the closer $(x, y, \ldots, z)$ is to $(x_0, y_0, \ldots, z_0)$.

For this reason, the tangent plane is used frequently as a local replacement for the function. In Chap. XIII, for example, it serves as the basis for Newton's method when solving two equations in two unknowns. As another example, we discuss now sensitivity analysis. We discussed this topic at the end of Sect. 1 of Chap. IV.

In calculations, the evaluation of a function f at a point $(x_0, y_0, \ldots, z_0)$ fails to be perfect for several reasons. One of these is the fact that, because of the limited number of digits carried in the calculator, or because of some earlier rounding error, the values in the calculator are not the numbers $x_0, y_0, \ldots, z_0$, but rather some perturbations $x_0 + \Delta x$, $y_0 + \Delta y, \ldots, z_0 + \Delta z$ of these numbers. This means that, even if all goes well from then on, we would be calculating the number $f(x_0 + \Delta x, y_0 + \Delta y, \ldots, z_0 + \Delta z)$ instead of the number $f(x_0, y_0, \ldots, z_0)$. Just how big an error are we making? The tangent plane gives an approximation for the difference,

$$f(x_0 + \Delta x, y_0 + \Delta y, \ldots, z_0 + \Delta z) - f(x_0, y_0, \ldots, z_0) \cong f_x \cdot \Delta x + f_y \cdot \Delta y + \ldots + f_z \cdot \Delta z , \tag{2.2}$$

with the partial derivatives all evaluated at $x_0, y_0, \ldots, z_0$. In fact, it can be shown that (2.2) can be made into an equality if we are prepared to evaluate all the partial derivatives at some unknown point of the form

$$(x_0 + \theta \Delta x,\ y_0 + \theta \Delta y, \ldots, z_0 + \theta \Delta z) ,$$

for some θ between 0 and 1. This means that if we have an indication of the size of the first partial derivatives $f_x, f_y, \ldots, f_z$ "near" $(x_0, y_0, \ldots, z_0)$, then we can estimate from (2.2) the effect of an uncertainty $\Delta x, \Delta y, \ldots, \Delta z$ in the arguments $x, y, \ldots, z$, respectively, on the accuracy of the calculated value $f(x_0 + \Delta x, y_0 + \Delta y, \ldots, z_0 + \Delta z)$ as an approximation to $f(x_0, y_0, \ldots, z_0)$.

Problem 2.1. Use (2.2) to estimate in % the possible combined effect in the calculated function value of a 0.01% error in each of the arguments, if f is as described below and the arguments are "near" the point given below.

(a) $f(x, y) = (x + y)/(x - y)$ $(0.9, 1.1)$

(b) $f(x, y) = x^y$ $(10, 10)$

(c) $f(x, y, z) = x^{y^z}$ $(100, 10, 1)$

(d) $f(x, y) = x \cos y$ $(100, 100)$

Also, determine in each of these cases whether the function is particularly sensitive to one of the variables.

Problem 2.2. It is stated in Example 24 of Sect. 1-6 of T-F that if the manager of a retail store is going to order a number Q of items to put into stock, then the best amount to ask for in a given order is

$$Q = \sqrt{\frac{2KM}{h}} , \tag{2.3}$$

where

K = cost of placing the order ,

M = number of items sold per week,

h = holding cost for each item per week.

(a) Write an approximation for ΔQ, with partial derivatives evaluated at $K_0 =$ \$2.00, $M_0 = 20$ radios per week, and $h_0 =$ \$0.05.

(b) At the values for K_0, M_0, h_0 given in (a), to which variable is Q most sensitive? least sensitive?

If we take $\Delta x = dx$, $\Delta y = dy$, $\Delta z = dz$, then the right side of (2.2) is called the differential of f; see Eq. (2) in Sect. 13-8 of T-F.

Problem 2.3. Using differentials to approximate increments, find approximately the amount of material in a hollow rectangular box whose inside measurements are 5 feet long, 3 feet wide, and 2 feet deep, if the box is made of lumber that is $\frac{1}{2}$ inch thick and the box has no top.

Remark. This is Prob. 2 at the end of Sect. 13-8 of T-F.

Problem 2.4. Find the amount of material in the box of Prob. 2.3 by getting the volume of the exterior of the box and subtracting the volume of the interior.

Note how much less calculation is required for Prob. 2.4 than for Prob. 2.3. Besides which, you get a much more accurate answer in Prob. 2.4. As we saw in Chap. VI, with the calculating power of your calculator, it is often easier to do an accurate calculation than to approximate by differentials.

Chapter XIII

MAXIMA AND MINIMA OF A FUNCTION OF SEVERAL VARIABLES

0. Guide for the reader.

If your calculus text does not cover this topic, you need not read this chapter. Otherwise, read the chapter when you get to that topic in your calculus course.

A specially simple case of minimization is finding the straight line which comes closest to fitting a set of data, in the sense of least squares. This is covered in Sect. 2. Some calculus texts do not treat this special case. We have made Sect. 2 independent of the rest of the chapter, so that it can be omitted in case it is not covered in your calculus course.

1. Critical points.

At a maximum or a minimum of a smooth function f, which is not at the boundary of f, all first partial derivatives must be zero. For this reason, it is customary when looking for a maximum or minimum of a function to look for points at which all the first partial derivatives of the function vanish. Such points are called critical points of the function f. Once all the critical points of f in a region are found, one determines its maximum or its minimum by maximizing or minimizing f only over these points and, of course, over the boundary points of the region in which the maximum or minimum of f is sought.

For a function f of two variables, you would have to solve the two simultaneous equations

$$\begin{aligned} f_x(x, y) &= 0 \\ f_y(x, y) &= 0 \end{aligned} \tag{1.1}$$

in the two unknowns x and y, in order to locate a critical point (x, y). Your calculus book is handicapped here, since it can use only examples for which it is pretty obvious how to solve the equations (1.1). But, with a little effort and a calculator, the number of examples for which you can find critical points (to within calculator accuracy) increases tremendously.

As a first step in such a calculation of critical points, you must locate such a critical point approximately. This you might do by plotting the function, as discussed at length in Sect. 1 of Chap. XII. An alternative is to sketch the two curves whose equations are

$$f_x(x, y) = 0, \qquad f_y(x, y) = 0,$$

respectively, and thereby locate, at least approximately, their intersections. These intersections are, of course, the critical points we are looking for.

For example, in Sect. 1 of Chap. XII, we made a sketch of the function

$$f(x, y) = x^3/3 + x(y^2 - 1) - y/2 + 1 \tag{1. 2}$$

and found that it has two critical points in the rectangle

$$R: \ -0.2 \leq x \leq 1.2, \qquad 0 \leq y \leq 1.2,$$

a minimum near (0. 95, 0. 25) and a saddle point near (0. 3, 0. 9). Consider now the two curves whose equations are, respectively, $f_x(x, y) = 0$ and $f_y(x, y) = 0$, for this example. We have

$$f_x(x, y) = x^2 + y^2 - 1, \qquad f_y(x, y) = 2xy - 1/2.$$

The corresponding curves are sketched in Fig. 1. 1. The first is a circle of radius 1 and with center at the origin, the other is a hyperbola, with the x-axis and the y-axis as its asymptotes. These curves intersect in four points, which are approximately the points

$$A(0.97,\ 0.26), \qquad B(0.26,\ 0.97)$$

$$C(-0.97, -0.26), \qquad D(-0.26,\ -0.97)$$

Our approximate determination of A, based on Fig. 1. 3 in Chap. XII, was really quite good!

You can actually even determine what kind of critical point each of these four points is, by paying attention to the directions of the gradients of f at various points. To recall, the gradient of f at a point (x, y) is the vector

$$\text{grad } f = f_x(x, y)\,\vec{i} + f_y(x, y)\,\vec{j}. \tag{1. 3}$$

This vector points in the direction of greatest growth of the function at the point (x, y). So, if the function has a minimum at some critical point, we would expect all nearby gradients to point away from that point. If it is a maximum, we would expect all nearby

Figure 1. 1

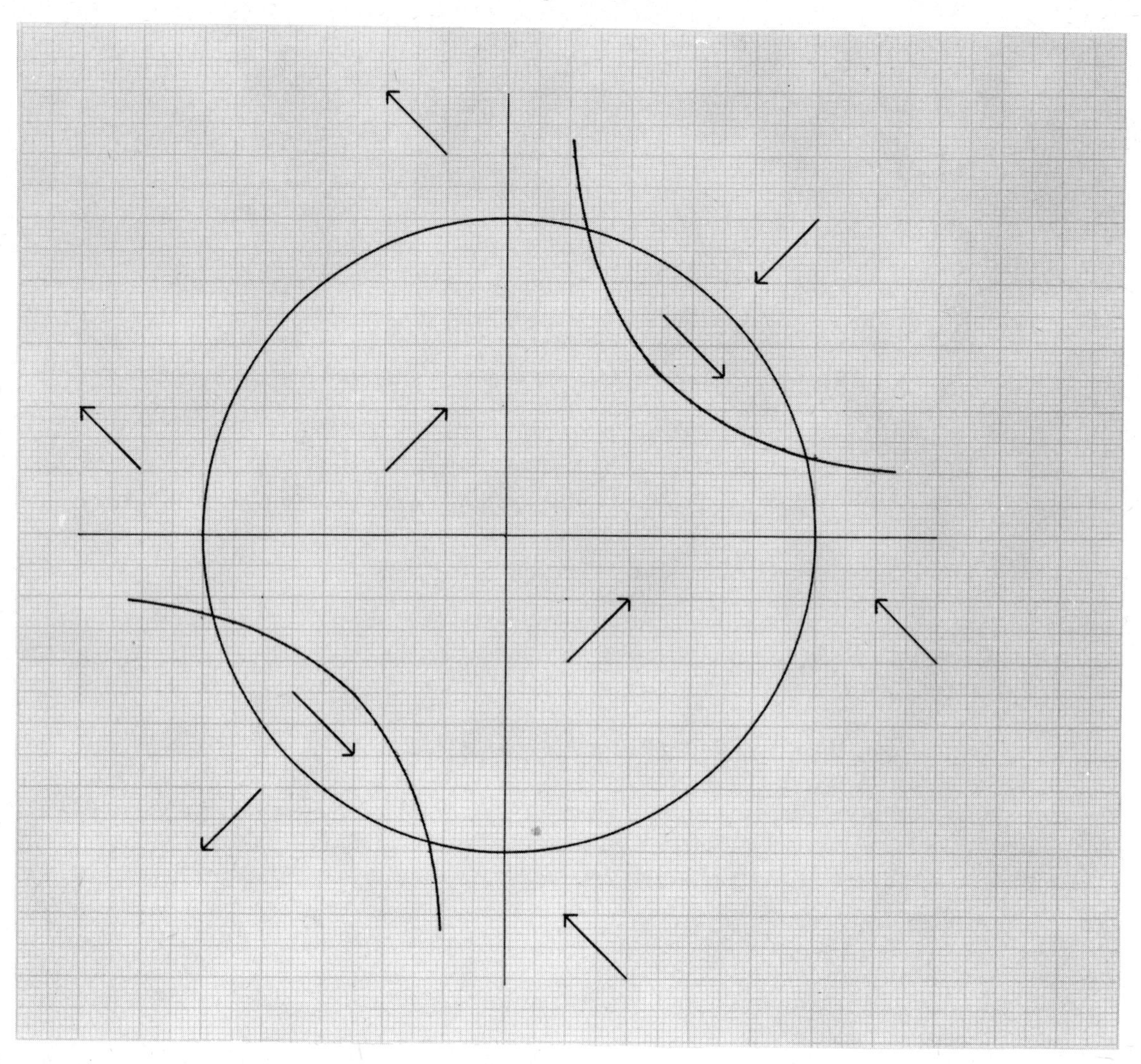

gradients to point more or less toward that point. And if some nearby gradients point toward a critical point and others point away from it, we would be dealing with a saddle point.

Now, once you have sketched the curves whose equations are $f_x(x, y) = 0$ and $f_y(x, y) = 0$, respectively, then you also know the regions where f_x is positive and where f_x is negative. Similarly, you know the regions where f_y is positive and those where f_y is negative. Since the numbers $f_x(x, y)$ and $f_y(x, y)$ form the components of the gradient, you are therefore aware of the general direction into which the gradient points in these regions.

For example, in Fig. 1. 1, $f_x < 0$ inside the circle and $f_x > 0$ outside that circle.

Again, $f_y < 0$ between the two branches of the hyperbola and $f_y >$ outside. So, in particular, grad f has both components negative in the region which is both inside the circle and between the two branches of the hyperbola. We have indicated this state of affairs in Fig. 1.1 by drawing into that region a vector in the direction of $-\vec{i} - \vec{j}$. This symbolizes the fact that, in that region, <u>all</u> gradients must point toward the third quadrant.

In the region inside the circle, but to the right of the hyperbola, $f_x < 0$ but $f_y > 0$. So we have drawn there a vector in the direction of the vector $-\vec{i} + \vec{j}$. This symbolizes that, in that region, all gradients must point toward the second quadrant.

We have proceeded similarly for the other regions of the figure.

With this information, a quick look at Fig. 1.1 indicates that A(0.97, 0.26) must be a minimum, C(-0.97, -0.26) must be a maximum, while the other two critical points must be saddle points.

<u>Problem 1.1.</u> For the surface

$$z = f(x, y) = 2xy - 5x^2 - 2y^2 + 4x + 4y - 4 , \tag{1.3}$$

sketch the curves given by $f_x(x, y) = 0$ and $f_y(x, y) = 0$, respectively. Find and classify all critical points, using the general gradient directions at nearby points.

<u>Problem 1.2.</u> Sketch the surface

$$z = f(x, y) = \sqrt{x^2 + y^2} \tag{1.4}$$

over the region R: $|x| \leq 1$, $|y| \leq 1$. Find the high and low points of the surface over R. Discuss the existence, and the values, of $\partial z/\partial x$ and $\partial z/\partial y$ at these points.

<u>Remark.</u> The two previous problems are Problems 4 and 7 at the end of Sect. 13-9 of T-F.

<u>Problem 1.3.</u> Sketch the curves given by $f_x(x, y) = 0$ and $f_y(x, y) = 0$, respectively, for the function f of Prob. 1.3 in Chap. XII, namely

$$f(x, y) = \frac{\sin(x + y)}{\cos(x - y) + 1.1} \qquad 1 < x < 3, \quad -2 \leq y \leq 0$$

Find and classify all critical points, using the general gradient directions at nearby points.

<u>Hint.</u> Both f_x and f_y are rational functions, and hence vanish exactly when their numerators vanish. Therefore, you can come up with pretty simple equations for these curves, particularly if you remember that $\cos(\alpha \pm \beta) = \cos\alpha \cos\beta \mp \sin\alpha \sin\beta$.

2. Least squares approximation to data by a straight line.

We follow T-F and study least squares approximation by a straight line as a simple application of how to find maxima and minima.

The problem is this. You are given some data points (x_n, y_n), $n = 1, \ldots, N$ in the plane and wish to determine a straight line

$$y = m^* x + b^* \tag{2.1}$$

which fits the data in the best possible way.

It is not at all clear how one should define what is meant by "best" here, or what is meant by "fitting". Of the many possibilities, we adopt here the "least squares" criterion which says that the straight line

$$y = mx + b$$

is <u>best</u> for which the <u>mean-square error</u>

$$E(m, b) = \sum_{n=1}^{N} [y_n - (mx_n + b)]^2 \tag{2.2}$$

is <u>as small as possible</u>. For, as it turns out, this straight line is particularly easy to determine.

Determination of this "best" line requires you to find a critical point (m^*, b^*) of the function E given by (2.2). Such a critical point satisfies the equations

$$E_m(m^*, b^*) = 0$$

$$E_b(m^*, b^*) = 0 .$$

We now look at these equations carefully.

Consider first E_b. Using the chain rule, we find that

$$E_b(m, b) = \sum_{n=1}^{N} 2[y_n - (mx_n + b)](-1)$$

$$= -2\{\sum_{n=1}^{N} y_n - m\sum_{n=1}^{N} x_n - \sum_{n=1}^{N} b\} .$$

It is convenient to use here the abbreviation

2. Least squares straight line approximation

$$\Sigma x = \sum_{n=1}^{N} x_n, \text{ hence } \Sigma y = \sum_{n=1}^{N} y_n ,$$

in terms of which the formula for $E_b(m, b)$ reads

$$E_b(m, b) = -2\{\Sigma y - m\Sigma x - Nb\} .$$

This shows that the equation $E_b(m^*, b^*) = 0$ is equivalent to the equation

$$\Sigma y - m^* \Sigma x - Nb^* = 0 \tag{2.3}$$

from which we get

$$b^* = [\Sigma y - m^* \Sigma x]/N. \tag{2.4}$$

Next, we look at E_m. We get from the chain rule that

$$\begin{aligned} E_m(m, b) &= \sum_{n=1}^{N} 2[y_n - (mx_n + b)](-x_n) \\ &= -2\{\sum_{n=1}^{N} y_n x_n - m\sum_{n=1}^{N} x_n^2 - b\sum_{n=1}^{N} x_n\} . \end{aligned}$$

With the further abbreviation

$$\Sigma xy = \sum_{n=1}^{N} x_n y_n, \text{ hence } \Sigma xx = \sum_{n=1}^{N} x_n^2 ,$$

this reads

$$E_m(m, b) = -2\{\Sigma xy - m\Sigma xx - b\Sigma x\}$$

and shows that the equation $E_m(m^*, b^*) = 0$ is equivalent to the equation

$$\Sigma xy - m^* \Sigma xx - b^* \Sigma x = 0 . \tag{2.5}$$

From this equation, we eliminate the $b^* \Sigma x$ term by multiplying the equation by N and then subtracting from it Σx times equation (2.3). This gives the new equation

$$N\Sigma xy - m^* N\Sigma xx - \Sigma x(\Sigma y - m^* \Sigma x) = 0$$

from which we obtain a formula for m^*,

$$m^* = \frac{N\Sigma xy - \Sigma x \cdot \Sigma y}{N\Sigma xx - \Sigma x \cdot \Sigma x} . \tag{2.6}$$

For easy reference, here is an algorithmic description of how to obtain the slope m^* and the y-intercept b^* of the straight line

$$y = m^*x + b^*$$

which best fits (in the least square sense) certain data points $(x_1, y_1), \ldots, (x_N, y_N)$.

<u>Step 1.</u> Accumulate the four sums

$$\Sigma x = \sum_{n=1}^{N} x_n, \quad \Sigma y = \sum_{n=1}^{N} y_n, \quad \Sigma xy = \sum_{n=1}^{N} x_n y_n, \quad \text{and} \quad \Sigma xx = \sum_{n=1}^{N} x_n^2 .$$

<u>Step 2.</u> Calculate $m^* = (N\Sigma xy - \Sigma x \cdot \Sigma y)/(N\Sigma xx - \Sigma x \cdot \Sigma x)$.

<u>Step 3.</u> Calculate $b^* = (\Sigma y - m^* \Sigma x)/N$.

<u>Remark.</u> The better programmable calculators have special keys which help materially in carrying out Step 1. For each n, you input x_n and y_n and then press a special key, commonly denoted by Σ+, and the calculator then adds $x_n, y_n, x_n y_n$, and x_n^2 (and even y_n^2) into certain registers, and adds 1 into a certain register in order to keep track of the number of points. You are, of course, expected to put zeros initially into each of these registers. At the end, you have all the information needed for Steps 2 and 3 (including the number N) at your fingertips. There is even a special key for correcting a mistake made when inputting one of the points (x_n, y_n). Consult your manual.

You will have to work a little harder if your calculator does not have that special Σ+ key. Your task is made more difficult by the fact that your calculator then probably has no more than just one memory register. In such a case, there is nothing to be done but for you to add up each of the sums Σx, Σy, Σxy, and Σxx. Recall that the M+ key will accumulate a sum in the one memory register, so that you can collect <u>two</u> sums at a time, for example the sum Σx in memory and the sum Σxx in the display, and also the sum Σy in memory and the sum Σxy in the display. For example, in order to form Σx and Σxx simultaneously, the typical steps, repeated N times, would be

RPN
|
RPN

x_n, [M+], [x^2], [+] .

AE
|
AE

[+], x_n, [M+], [x^2] .

<u>Problem 2. 1.</u> Construct the least squares straight line fit to data (x_n, y_n), $n = 1, \ldots, 10$ obtained as follows. $x_n = n$, and y_n is the value at x_n of the linear function f given by

$$f(x) = \pi x - e ,$$

but rounded to the nearest integer. Compare the coefficients m^* and b^* you get in this way with the coefficients π and $-e$ of the original line.

Note. Here, e is the base for the natural logarithm (obtainable on your calculator by evaluating e^x or $\ln^{-1}x$ at $x = 1$). Also, an easy way to round a nonnegative number to the nearest integer is to add 0.5 to it and then truncate it, i. e., take the integer part of the result. Many calculators have an INT key for taking the integer part. If you are dealing with a negative number, you would have to subtract 0.5 from it and then use the INT key.

Problem 2. 2. A scientist has measured, at various temperatures, the pressure in a gas filled container which is being slowly heated. Her measurements are given in the following table.

temperature $^{\circ}$C	35	24	36	49	74	100
pressure kg/cm^2	81	78	81	84	91	98

Calculate the average increase in pressure per degree Celsius increase in temperature as the slope of the least squares straight line fit to these data.

Problem 2. 3. Determine a formula for the least squares fit

$$y = b^*$$

to given data (x_n, y_n), $n = 1, \ldots, N$, by a constant function. What is the number b^* called in this case?

Problem 2. 4. Write a linear equation for the effect of irrigation on the yield of alfalfa by fitting a least-squares straight line to the following data from the University of California Experimental Station, Bulletin No. 450, p. 8. Plot the data and draw the line.

Table 2. 1

x (total seasonal depth of water applied (inches))	12	18	24	30	36	42
y (average alfalfa yield (tons/acre))	5.27	5.68	6.25	7.21	8.20	8.71

Remark. This is Prob. 6 at the end of Sect. 13-10 of T-F.

3. The solution of a linear system of two equations in two unknowns.

For future developments, we need to know how to solve the pair of equations

$$a_{11}x + a_{12}y = b_1 \tag{3. 1}$$

$$a_{21}x + a_{22}y = b_2 \tag{3. 2}$$

for the values of the two unknowns x and y. Many calculus texts suggest Cramer's rule as a means of finding the solution. For use on a calculator, the method of elimination is considerably superior, and we shall now explain it. In fact, a calculator program for Cramer's rule takes from 40% to 80% more steps (depending on the kind of calculator) than one for elimination. If you doubt this, try writing a program for Cramer's rule and compare it with the program for elimination, which we shall give below.

The process of elimination goes as follows. Divide (3. 1) through by a_{11} (assuming that $a_{11} \neq 0$) and get a new equation

$$x + a'_{12}y = b'_1 \tag{3. 3}$$

with

$$a'_{12} = a_{12}/a_{11}, \quad b'_1 = b_1/a_{11} .$$

Multiply (3. 3) through by a_{21} and subtract it from (3. 2) in order to eliminate x. This gives

$$a'_{22}y = b'_2 \tag{3. 4}$$

with

$$a'_{22} = a_{22} - a_{21}a'_{12}, \quad b'_2 = b_2 - a_{21}b'_1 .$$

Now (assuming that $a'_{22} \neq 0$), solve (3. 4) for y, getting

$$y = b'_2 / a'_{22} . \tag{3. 5}$$

Substitute this into (3. 3), getting

$$x = b'_1 - a'_{12}y . \tag{3. 6}$$

To do this by a calculator program, you first put $a_{11}, a_{12}, b_1, a_{21}, a_{22}$, and b_2 in registers one through six, respectively. Then use the following program, which will put the solutions x and y into registers three and six respectively.

RPN

(3.7) [RCL 1], [STO ÷ 2], [STO ÷ 3],
[RCL 2], [RCL 4], [×], [STO − 5],
[RCL 3], [RCL 4], [×], [STO − 6],
[RCL 5], [STO ÷ 6],
[RCL 6], [RCL 2], [×], [STO − 3].

This takes 17 program steps.

RPN

AE

(3.7) [RCL 1], [STO ÷ 2], [STO ÷ 3],
[RCL 2], [×], [RCL 4], [=], [STO − 5],
[RCL 3], [×], [RCL 4], [=], [STO − 6],
[RCL 5], [STO ÷ 6],
[RCL 6], [×], [RCL 2], [=], [STO − 3].

This takes a total of 20 program steps.

AE

Problem 3.1. Solve each of the following linear systems using Program (3.7). Explain any error messages you receive from your calculator.

(a) $\begin{aligned} 20x + 60y &= 78 \\ 10x + 20y &= 30 \end{aligned}$ (b) $\begin{aligned} 0x + 60y &= 78 \\ 10x + 20y &= 30 \end{aligned}$ (c) $\begin{aligned} 30x + 60y &= 78 \\ 10x + 20y &= 30 \end{aligned}$

What should you do with a system like (b) so that Program (3.7) will handle it successfully?

Once you have solved the linear system (3.1) and (3.2) by Program (3.7), you can save some inputting and some program steps in case you then have to solve another linear system for which only the right hand sides are different. In such a case, you need not reload a_{11}, a_{12}, a_{21} and a_{22}. You need only load the new b_1 and b_2 into registers three and six, respectively, and then omit the second step and the second line from Program (3.7).

Problem 3.2. Use this procedure to solve the linear systems

(a) $\begin{aligned} 20x + 60y &= 74 \\ 10x + 20y &= 24 \end{aligned}$ (b) $\begin{aligned} 20x + 60y &= 1 \\ 10x + 20y &= 0 \end{aligned}$ (c) $\begin{aligned} 20x + 60y &= 0 \\ 10x + 20y &= 1 \end{aligned}$

4. Newton's method.

In this section, we present Newton's method for solving two equations in two unknowns. See Prob. 7 at the end of Sect. 13-8 of T-F. The occasion for this discussion is, of course, our attempt in this chapter to find critical points of a function of two variables, which requires us to solve the simultaneous equations

$$(4.1)\qquad \begin{aligned} f_x(x,y) &= 0 \\ f_y(x,y) &= 0 . \end{aligned}$$

But, the method is applicable to any system

$$(4.2)\qquad \begin{aligned} g(x,y) &= 0 \\ h(x,y) &= 0 \end{aligned}$$

of two equations in two unknowns, and we will therefore discuss it in these general terms.

The basic idea of Newton's method is as follows. Take a guess (x_0, y_0) for a solution (x, y) of (4. 2) and replace both g and h in (4. 2) by their tangent planes at the point (x_0, y_0). This gives the system

$$(4.3)\qquad \begin{aligned} g(x_0,y_0) + g_x(x_0,y_0)(x - x_0) + g_y(x_0,y_0)(y - y_0) &= 0 \\ h(x_0,y_0) + h_x(x_0,y_0)(x - x_0) + h_y(x_0,y_0)(y - y_0) &= 0 \end{aligned}$$

which is linear in the unknowns x and y, so can be solved by Program (3. 7).

Actually, it is more convenient to think of (4. 3) as a linear system for the corrections $\delta x = x - x_0$, $\delta y = y - y_0$,

$$(4.4)\qquad \begin{aligned} g_x\delta_x + g_y\delta_y &= -g \\ h_x\delta_x + h_y\delta_y &= -h \end{aligned}$$

all functions to be evaluated at (x_0, y_0). This you would solve for the corrections δx and δy, using Program (3. 7), and then obtain a new guess

$$(4.5)\qquad (x_1, y_1) = (x_0 + \delta x,\ y_0 + \delta y).$$

If (x_0, y_0) is "close enough" to a solution of (4. 2), then (x_1, y_1) will be even closer and a few repetitions of this process will usually procure a solution of (4. 2) to as many digits as your calculator carries.

An example might be helpful. We found in Sect. 1 that the function f given by

$$(4.6)\qquad f(x,y) = \frac{x^3}{3} + x(y^2 - 1) - \frac{y}{2} - 8$$

has a local minimum near the point $(1, 0)$. Since $f_x(x,y) = x^2 + y^2 - 1$, $f_y(x,y) = 2xy - 0.5$, the equations which this (and any other) critical point of f must satisfy are of the form (4.2) with

$$g(x, y) = x^2 + y^2 - 1$$

$$h(x, y) = 2xy - 0.5$$

Since $g_x(x, y) = 2x$, $g_y(x, y) = 2y$, $h_x(x, y) = 2y$, $h_y(x, y) = 2x$, the Newton equations (4.4) for this example take the form

$$(4.7) \quad \begin{aligned} (2x_0)\delta x + (2y_0)\delta y &= -(x_0^2 + y_0^2 - 1) \\ (2y_0)\delta x + (2x_0)\delta y &= -(2x_0 y_0 - 0.5) \end{aligned}$$

At $(x_0, y_0) = (1, 0)$, this simplifies to

$$\begin{aligned} 2\delta x \quad &= 0 \\ 2\delta y &= 0.5 \end{aligned}$$

and so gives the new guess

$$(x_1, y_1) = (x_0 + \delta x,\ y_0 + \delta y) = (1 + 0,\ 0 + 0.25) = (1, 0.25) .$$

We try to improve this guess further. On replacing (x_0, y_0) in the Newton equations (4.7) by (x_1, y_1), we get the new system

$$\begin{aligned} 2\,\delta x + 0.5\,\delta y &= -0.0625 \\ 0.5\,\delta x + \quad 2\,\delta y &= \quad 0 \end{aligned}$$

and from this we find $\delta x = -0.03333\ 33333$, $\delta y = 0.00833\ 33333$, hence

$$\begin{aligned} (x_2, y_2) &= (x_1 + \delta x,\ y_1 + \delta y) \\ &= (0.96666\ 66667,\ 0.25833\ 33333) . \end{aligned}$$

We record these steps and further steps in Table 4.1, as calculated on an HP-33E.

Table 4.1

r	x_r	y_r	$g(x_r, y_r)$	$h(x_r, y_r)$
0	1	0	0	-0.5
1	1	0.25	0.0625	0
2	0.96666 66667	0.25833 33333	0.00118 05560	-0.00055 55556
3	0.96592 63703	0.25881 85275	0.00000 07830	-0.00000 07184
4	0.96592 58263	0.25881 90451	0	0

Table 4.1 shows the rapid convergence of Newton's method, with the proverbial "doubling" of the number of accurate digits quite visible.

For easy reference, we now give a concise algorithmic description of Newton's method for solving (4.2).

<u>Preparation.</u> Pick a guess (x_0, y_0) for a solution of $g(x,y) = 0$ and $h(x,y) = 0$.

<u>Loop.</u> Repeat for $r = 0, 1, \ldots,$ until terminated at Step 4:

<u>Step 1.</u> Evaluate $g, g_x, g_y, h, h_x,$ and h_y, all at the current guess (x_r, y_r). Form the Newton equations

$$g_x \delta x + g_y \delta_y = -g$$

$$h_x \delta x + h_y \delta_y = -h$$

for the correction $(\delta x, \delta y)$.

<u>Step 2.</u> Solve the Newton equations (e.g., by Program (3.7)).

<u>Step 3.</u> Calculate $x_{r+1} = x_r + \delta x$, and $y_{r+1} = y_r + \delta y$.

<u>Step 4.</u> Stop if g and h seem small enough. Otherwise, go back to Step 1, with $r+1$ in place of r.

Before attempting Prob. 4.1, below, read the discussion that follows it.

<u>Problem 4.1.</u> For each of the following functions, find a critical point near the specified point, to the accuracy of your calculator. Then determine whether it belongs to a local maximum, a local minimum or neither.

(a) $x^4 - 3xy^2 + y^3$ $\quad$ (4, 7)

(b) $x^4 - 4xy^2 + y^3 - x - y$ $\quad$ (0.5, 0)

(c) $x^4 - 4xy^2 + y^3 - x - y$ $\quad$ (7, 19).

In the problem above, if you have a nonprogrammable calculator, stop when you think you have x and y correct to 5 significant digits each.

Unless you have one of the better programmable calculators, you may find it a bit inadequate for the problem above. For example, neither the HP-33E nor the TI-57 can quite accomodate the program for one iteration of the Newton algorithm given above. However, you can manage by doing a little bit of the iteration by hand and the rest by a stored program. Thus, for part (a), we have $g_x = 12x^2$, $g_y = -6y$, $h_x = -6y$, $h_y = -6x + 6y$. To calculate these, and store them in registers one, two, four, and five, respectively, is not a very laborious hand calculation. Then you can run a stored program which calculates $g(x, y)$ and $h(x, y)$, and stores them in registers three and six, then executes Program (3. 7), and finally adds δx to x and δy to y. Fortunately, registers zero and seven are available to hold x and y.

In part (c), there is a little problem with noise. Each of $g(x, y)$ and $h(x, y)$ involves a term larger than 1000. So, on a 10 digit calculator, roundoff error can run as high as two or three times 10^{-6}. However, if x and y are both within two or three units in the last place of their true values, the values of $g(x, y)$ and $h(x, y)$ are both less than 10^{-6} in absolute value. So attempts to calculate g and h will result in small multiples of 10^{-6}, which are composed almost entirely of roundoff, and have nothing whatsoever to do with the true values of g and h. At the end of Sect. 2 of Chap. IV, this phenomenon is called noise.

In this situation, the Newton algorithm cannot get you any closer to the true values of x and y than a few units in the last digit. When you get that close, attempts to calculate g and h will produce quite irrelevant values for the right side of the Newton equations, (4. 4). Then the solution of (4. 4), say by Program (3. 7), will produce quite irrelevant values of δx and δy. If you persist with more iterations of the algorithm, x and y will jump around aimlessly, not by very much.

So, you should just stop.

Let us look at a function very similar to the one defined in (4. 6), namely

$$f(x, y) = \frac{x^3}{3} + x(y^2 - 1) - y - 8 . \tag{4. 8}$$

For this function, the curves $f_x(x, y) = 0$ and $f_y(x, y) = 0$ are tangent to each other at the critical point. This means that as the trial point (x_r, y_r) gets nearer and nearer to the critical point, the lines (4. 3) get more and more parallel. This means that the solution of (4. 4) becomes more subject to error.

The result of all this is that the Newton algorithm converges very slowly, and runs into noise at an early stage. This is shown in Table 4. 2. Observe that quite early, the values of g and h settle into a pattern, where each is about a quarter of the one above. However, at $r = 14$, this pattern begins to break up, and from then on the values of g and h are only noise.

What we are trying to solve are

$$g(x, y) = x^2 + y^2 - 1 = 0 \tag{4. 9}$$

$$h(x, y) = 2xy - 1 = 0 . \tag{4. 10}$$

As suggested in Prob. 3. 9 of Chap. IV, let us put

$$x = 0.7071 + h, \qquad y = 0.07071 + k .$$

Table 4. 2

r	x_r	y_r	$10g(x_r, y_r)$	$10h(x_r, y_r)$
0	1	0	0	-10
1	1	0. 5	2. 5	0
2	0. 8333 33333	0. 5833 33333	0. 34722 22150	-0. 2777 77785
3	0. 7696 07843	0. 6446 07843	0. 07815 50325	-0. 0780 94965
4	0. 7383 56782	0. 6758 56781	0. 01953 12600	-0. 0195 31240
5	0. 7227 31782	0. 6914 81780	0. 00488 07500	-0. 0048 82820
6	0. 7149 19281	0. 6992 94282	0. 00122 07100	-0. 0012 20695
7	0. 7110 13032	0. 7032 00530	0. 00030 51725	-0. 0003 05180
8	0. 7090 59905	0. 7051 53658	0. 00007 63050	-0. 0000 76285
9	0. 7080 83328	0. 7061 30234	0. 00001 90675	-0. 0000 19080
10	0. 7075 95032	0. 7066 18530	0. 00000 47625	-0. 0000 04775
11	0. 7073 50857	0. 7068 62705	0. 00000 11875	-0. 0000 01195
12	0. 7072 28841	0. 7069 84721	0. 00000 02925	-0. 0000 00305
13	0. 7071 67652	0. 7070 45910	0. 00000 00675	-0. 0000 00080
14	0. 7071 37363	0. 7070 76200	0. 00000 00275	-0. 0000 00010
15	0. 7071 22035	0. 7070 91528	0. 00000 00150	+0. 0000 00005
16	0. 7071 13840	0. 7070 99723	0. 00000 00100	+0. 0000 00010
17	0. 7071 13840	0. 7070 99723	0. 00000 00100	+0. 0000 00010

Then (4. 9) and (4. 10) become

(4. 11) $$h^2 + 1.4142h + k^2 + 1.4142k - 9.59 \times 10^{-5} = 0$$

(4. 12) $$2hk + 1.4142h + 1.4142k - 1.918 \times 10^{-4} = 0$$

Problem 4. 2. Find a solution (h, k) of (4. 11) and (4. 12) that is near $(0, 0)$.

Remark. You will encounter noise again, and perhaps not get values of h and k to more than four or five significant digits. However, adding these to 0. 7071 will give you x and y to 8 or 9 significant digits.

By then (indeed considerably earlier), it should occur to you that x and y are suspiciously near to being equal. It can't hurt to put $y = x$ in (4. 9) and (4. 10), and see what happens. When you do, you get the exact solution

$$y = x = \frac{\sqrt{2}}{2} .$$

What have we said about being alert? You could have tried this fairly early and saved yourself a lot of work.

Problem 4. 3. Find a critical point of

$$x^4 - 3xy^2 + y^3$$

near $(0.2, 0.1)$.

Remark. This converges very slowly indeed. So you possibly wasted a lot of time before it dawned on you to try to see if $(0, 0)$ is the critical point in question, which it obviously is.

5. Damped Newton's method.

As with Newton's method for one equation in one unknown, Newton's method for a system usually converges quadratically when it converges. This means the following. If we denote by $\vec{P}_n$ the vector from the origin to our n-th guess $P(x_n, y_n)$, and by $\vec{P}$ the vector from the origin to the solution $P(x, y)$ of (4.2), then

$$(5.1) \qquad |\vec{P} - \vec{P}_{n+1}| \approx \text{const}\ |\vec{P} - \vec{P}_n|^2 .$$

An immediate consequence of such quadratic convergence is the fact that the ratio of successive corrections goes to zero,

$$(5.2) \qquad \frac{|\vec{P}_{n+1} - \vec{P}_n|}{|\vec{P}_n - \vec{P}_{n-1}|} \xrightarrow[n\to\infty]{} 0$$

If this does not happen, but rather

$$(5.3) \qquad \frac{|\vec{P}_{n+1} - \vec{P}_n|}{|\vec{P}_n - \vec{P}_{n-1}|} \xrightarrow[n\to\infty]{} \text{const} \neq 0$$

then you are almost certain to have made a mistake in your calculation of the functions g_x, g_y, g, h_x, h_y, h as you were constructing the Newton equations (4.4). Since derivatives of functions are usually more complex than the functions themselves, such mistakes are easy to make and are made all the time by people using Newton's method. You have to watch out for them because they slow down convergence, and often prevent convergence altogether.

An even more disagreeable hazard in the use of Newton's method is the requirement that the initial guess (x_0, y_0) be "close enough" to a solution of (4.3) in order to get convergence at all. It is clear that (x_0, y_0) must be chosen with some care, since its choice will influence to which solution of (4.3) the iteration process converges (if it does). But making a good choice often requires some elaborate sketching of the two curves

$$g(x, y) = 0 \quad \text{and} \quad h(x, y) = 0$$

with a view toward determining their intersections graphically to sufficient accuracy for

Newton's method to work.

Often such labor can be avoided by the use of the damped Newton's method. In this method, you monitor the size of the error function E given by

$$E(x, y) = g(x, y)^2 + h(x, y)^2 \tag{5.4}$$

as you go along. This means that you elaborate on Step 3 of the algorithm given in Sect. 4, as follows.

Step 3.1. Let $\Delta x = \delta x$, $\Delta y = \delta y$.

Step 3.2. Calculate $(x_{n+1}, y_{n+1}) = (x_n + \Delta x, y_n + \Delta y)$.

Step 3.3. Now check whether

$$E(x_{n+1}, y_{n+1}) < E(x_n, y_n) .$$

If it is, go on to Step 4. Otherwise, cut Δx and Δy in half, i.e., replace Δx by $\Delta x/2$ and replace Δy by $\Delta y/2$, and go back to Step 3.2.

In this way, you are assured that the number $E(x_n, y_n)$ decreases for each n. This is a good thing since you are looking ultimately for some (x, y) for which $E(x,y) = 0$.

You may (and, in any event, you should) wonder whether you'll ever get out of that loop formed by Steps 3.2 - 3.3. Perhaps, no matter how many times you cut Δx and Δy in half, you'll always get

$$E(x_n + \Delta x, y_n + \Delta y) > E(x_n, y_n)?$$

According to calculus, this cannot be. For, each of the points $(x_n + \Delta x, y_n + \Delta y)$ is of the form

$$(x_n + t\delta x, y_n + t\delta y)$$

for some t of the form 2^{-k}, and it would therefore follow that the function η given by

$$\eta(t) = E(x_n + t\delta x, y_n + t\delta y)$$

has $\eta'(0) \geq 0$. On the other hand, you can actually calculate $\eta'(0)$ since it is the directional derivative of E at (x_n, y_n) in the direction of the vector

$$\vec{\delta} = (\delta x)\vec{i} + (\delta y)\vec{j} ,$$

- up to a positive factor. Precisely, you know from calculus that

$$\eta'(0) = (\text{grad } E) \cdot \vec{\delta} .$$

You can also figure out that

$$E_x = 2gg_x + 2hh_x, \qquad E_y = 2hh_x + 2hh_y .$$

Putting these two facts together, you get

$$\eta'(0) = 2[(gg_x + hh_x)(\delta x) + (gg_y + hh_y)(\delta y)]$$

$$= 2[g(g_x \delta x + g_y \delta y) + h(h_x \delta x + h_y \delta y)] .$$

But now, since $(\delta x, \delta y)$ solves the Newton equations (4. 4), this last expression equals

$$2[g(-g) + h(-h)] = -2E$$

which is <u>negative</u> (unless $g = h = 0$). This shows that

$$E(x_n + t\delta x, \; y_n + t\delta y) < E(x_n, y_n)$$

for all positive t near 0 and therefore guarantees that the loop formed by Steps 3. 2 - 3. 3 eventually terminates.

Of course, if you find yourself going through it too often for your taste, you might give up on your current guess and start from another, or else put the problem on a more powerful computer.

<u>Problem 5. 1</u>. Find a solution to the system

$$x + 3 \cdot \ln|x| - y^2 = 0$$

$$2x^2 - xy - 5x = -1$$

near the point $(2.5, 3.5)$.

<u>Problem 5. 2.</u> The system of equations in Prob. 5. 1 has a second solution, in the fourth quadrant. Find it.

<u>Remark</u>. Program XIII. 1 in the Program Appendix gives an outline for the damped Newton's method.

Chapter XIV

SERIES

0. Guide for the reader.

You should wait until you are well into the discussion of series in the calculus before reading the material of this chapter. For instance, wait until you have learned about the series for $\sin x$, $\cos x$, e^x, and perhaps $\ln(1 + x)$. Whether, and how, the calculator can help with the various calculations that arise in connection with series is a complex question, and is discussed in Sect. 1. In Sect. 2, we get off into the calculation of functions that cannot be keyed in on the calculator. Often one can approximate these by series, and illustrations are given.

1. Calculation of series by calculator.

A famous series is

$$e^x = 1 + \frac{x}{1!} + \frac{x^2}{2!} + \frac{x^3}{3!} + \ldots \quad . \tag{1.1}$$

There are an infinite number of terms in the series, so that it is hopeless even to think of adding them all up. But one of the things you learn is that if you add up the first N terms, you get an approximation to e^x. The larger N is, the better the approximation. Indeed, calculus gives formulas which tell how large you have to take N to approach within ϵ of e^x; it depends on x, of course, but that is taken care of by the formula.

But if you take only N terms on the right of (1.1), you have a polynomial. A good way to calculate a value of a polynomial is by Horner's method. Specifically, if you stop with $x^N/N!$, you would calculate

$$(\ldots((\frac{x}{N!} + \frac{1}{(N-1)!})x + \frac{1}{(N-2)!})x + \ldots + \frac{1}{1!})x + 1 \ .$$

The calculation of $N!$, $(N-1)!$, etc., is a bother. You would do better to put the formula into the alternate form

(1. 2) $$((\ldots((\frac{x}{N}+1)\frac{x}{N-1}+1)\frac{x}{N-2}+\ldots+1)\frac{x}{2}+1)\frac{x}{1}+1 .$$

This is a very quick calculation, even if N is fairly large, especially on a programmable calculator.

But wait. The calculator has a key labelled e^x. Simply input x and press the e^x key, and out comes an approximation for e^x. No bother about how large one has to take N. No fuss about programming (1. 2).

Most of the series treated in calculus are for the calculation of functions for which we now have keys on the calculator. For example, e^x, $\sin x$, $\cos x$, $\ln(1+x)$, $\tan^{-1}x$, $(1+x)^m$, and so on. So, although the calculator could help in summing these series, why bother? Just press the key for e^x, $\sin x$, $\cos x$, $\ln(1+x)$, $\tan^{-1}x$, $(1+x)^m$, and so on. The same goes for the various series to calculate π. There is a π key on the calculator.

Before calculators, one had to use these series to calculate these various functions. And quite a labor it was. Of course people had calculated tables for numerous of them (using the series, of course), which helped a lot. But now we have calculators.

On the other hand, the calculator gives only an approximation for e^x. It is pretty accurate. But suppose that for some x you need a more accurate approximation for e^x than your calculator gives. Well, there is the series (1. 1). And your calculus text has a formula that says how large you have to take N to get the desired accuracy. See (2. 11) in Chap. IV. The same for $\sin x$, $\cos x$, and the rest.

<u>Problem 1. 1.</u> There is the famous series

(1. 3) $$\cos x = 1 - \frac{x^2}{2!} + \frac{x^4}{4!} - \frac{x^6}{6!} + \frac{x^8}{8!} - \ldots .$$

Suppose you stop at the term $(-1)^N x^{2N}/(2N)!$, thereby getting a polynomial approximation for $\cos x$. Analogous to (1. 2), specify $A_1, A_2, A_3, \ldots, A_N$ so that you can evaluate this polynomial by calculating

(1. 4) $$((\ldots((A_N x^2+1)A_{N-1}x^2+1)A_{N-2}x^2+\ldots+1)A_2x^2+1)A_1x^2+1 .$$

2. Series for functions that cannot be keyed in on the calculator.

On the whole, calculus texts avoid functions that cannot be keyed in on a calculator. However, since one can write series for some of them, a few sometimes appear in the discussion of series.

One such is defined by

(2. 1) $$f(x) = \int_0^x \frac{\sin y}{y}\,dy .$$

See Prob. 9 at the end of Sect. 16 - 10 of T - F.

From the result in Example 2 of Sect. 16 - 9 of T - F, we have

(2. 2) $$\frac{\sin y}{y} = 1 - \frac{y^2}{3!} + \frac{y^4}{5!} - \frac{y^6}{7!} + \dots + \frac{(-1)^N y^{2N}}{(2N+1)!}$$

with an error no greater than

(2. 3) $$\frac{|y|^{2N+1}}{(2N+2)!} .$$

Integrating both sides from 0 to x gives

(2. 4) $$f(x) = x - \frac{x^3}{3(3!)} + \frac{x^5}{5(5!)} + \dots + \frac{(-1)^N x^{2N+1}}{(2N+1)((2N+1)!)}$$

with an error no greater than

(2. 5) $$\frac{|x|^{2N+2}}{(2N+2)((2N+2)!)} .$$

The calculation of the polynomial on the right side of (2. 4) is no particular problem. For instance, if we take $N = 3$, the right side of (2. 4) can be written as

(2. 6) $$(((\frac{1}{7}\frac{(-x^2)}{6\cdot 7} + \frac{1}{5})\frac{(-x^2)}{4\cdot 5} + \frac{1}{3})\frac{(-x^2)}{2\cdot 3} + 1)\, x .$$

Problem 2. 1. Get a 5 decimal approximation for

(2. 7) $$\int_0^1 \frac{\sin y}{y}\, dy .$$

Remark. An accurate approximation for (2. 7) is

0. 94608 30704 .

Problem 2. 2. Derive the series

(2. 8) $$\int_0^x \sin y^2\, dy = \frac{x^3}{3} - \frac{x^7}{7(3!)} + \frac{x^{11}}{11(5!)} - \frac{x^{15}}{15(7!)} + \dots .$$

Remark. This is derived in Example 8 of Sect. 16 - 13 of T - F.

Problem 2. 3. Find the error if you stop with

2. Series for functions that cannot be keyed in

$$\frac{(-1)^N x^{4N+3}}{(4N+3)((2N+1)!)} \tag{2. 9}$$

on the right side of (2. 8).

Problem 2. 4. Find a formula like (2. 6) for evaluating the polynomial that would result from stopping at $-x^{15}/(15(7!))$ in the right side of (2. 8).

Problem 2. 5. Find to 5 decimals

$$\int_0^1 \sin y^2 \, dy . \tag{2. 10}$$

Remark. The function on the left side of (2. 8) is much used in the theory of optics. Thus it is quite useful to be able to calculate it. However, it cannot be calculated directly by any combination of calculator keys, so that it is good to have the series (2. 8) for it. Because the series (2. 8) is known, and an error term is known (see your answer to Prob. 2. 3), an ambitious calculator manufacturer could build a calculator with a key that would give an approximation for the left side of (2. 8). Possibly this will be done sometime, for special use by people working in optics.

Incidentally an accurate value for (2. 10) is

$$0.\,31026\;\;83017$$

as indicated in Example 9 of Sect. 16-13 of T-F. You can verify this by taking $x = 1$ in a formula like the one you got in Prob. 2. 4.

If you put $-y^2$ for x in (1. 1), and integrate from 0 to x, you get

$$\int_0^x e^{-y^2} dy = x - \frac{x^3}{3(1!)} + \frac{x^5}{5(2!)} - \frac{x^7}{7(3!)} + \dots . \tag{2. 11}$$

This could be used for Prob. 10 at the end of Sect. 16-10 of T-F. By working with the error term for (1. 1), you could derive an error term for stopping at a certain point on the right of (2. 11).

Problem 2. 6. Find a formula like (1. 2) for evaluating the polynomial that would result from stopping at $x^9/(9(4!))$ in the right side of (2. 11)

Remark. The function on the left side of (2. 11) is much used in the theory of statistics. This in turn, is considerably used in engineering and the various sciences.

By taking $m = \frac{1}{2}$ and $x = y^4$ in the equation

$$(1 + x)^m = 1 + mx + \frac{m(m-1)}{2!} x^2 + \dots + \frac{m(m-1)(m-2)\dots(m-k+1)}{k!} x^k + \dots$$

(see (8) of Sect. 16-8 of T-F), and integrating from 0 to x, we get

$$\int_0^x \sqrt{1+y^4}\,dy = x + \frac{1}{2}\frac{x^5}{5(1!)} + \left(\frac{1}{2}\right)\left(-\frac{1}{2}\right)\frac{x^9}{9(2!)} \tag{2.12}$$

$$+ \left(\frac{1}{2}\right)\left(-\frac{1}{2}\right)\left(-\frac{3}{2}\right)\frac{x^{13}}{13(3!)} + \left(\frac{1}{2}\right)\left(-\frac{1}{2}\right)\left(-\frac{3}{2}\right)\left(-\frac{5}{2}\right)\frac{x^{17}}{17(4!)} + \cdots .$$

This is like what T-F do in Example 10 in Sect. 16-13. Suppose we wish

$$\int_0^2 \sqrt{1+y^4}\,dy \quad . \tag{2.13}$$

Do we simply put $x = 2$ on the right side of (2.12)? This will not work, since the resulting series diverges, and no sensible sort of answer would be forthcoming. It is the case (though this is not very easy to prove) that the right side of (2.12) converges for $|x| \le 1$ and diverges for $|x| > 1$. So (2.12) works fine for Example 10 in Sect. 16-13 of T-F, where they take $x = 0.5$. However, for $x = 2$, as in (2.13), it is of no use whatsoever.

So series cannot be used for everything.

This does not mean that we have no way to get an approximate value for (2.13). There is always Simpson's rule for carrying out an approximate numerical integration. See Sect. 4 of Chap. X.

Incidentally, the terms shown on the right of (2.12) can be calculated very quickly by means of the formula

$$((((\frac{1}{17}\,\frac{5}{8}(-x^4) + \frac{1}{13})\,\frac{3}{6}(-x^4) + \frac{1}{9})\,\frac{1}{4}(-x^4) + \frac{1}{5})\,\frac{1}{2}x^4 + 1)\,x \ . \tag{2.14}$$

Chapter XV

DIFFERENTIAL EQUATIONS

0. Guide for the reader.

This chapter tells how to get numerical approximations for solutions of differential equations. Most calculus texts are carefully written so as to require very little calculation on the part of the reader. So the likelihood that you would be called on to use any of the methods of this chapter during the actual calculus course is very small (especially if your text does not even consider differential equations, as some do not). But, in most scientific or engineering endeavors, one does have to get such numerical approximations. If your interests lie in these directions, ignorance of the methods of this chapter would put you under a handicap.

But why is there a problem? As you read through the chapter in the calculus text, for each differential equation that they present they always come up with a solution

$$y = g(x) .$$

Just program your calculator to calculate $g(x)$. Then, every time somebody wishes an approximate value of y for some x, you simply input that x and press the program key (or keys).

It isn't always that easy. Consider, for example, the following very simple differential equation

$$\frac{dy}{dx} = f(x) .$$

It is so simple, it is often not even mentioned in chapters on differential equations. But it is a differential equation nevertheless, since it relates the value of a function and of its derivatives at each value of the independent variable. You have, in the chapters on integration, learned how to solve this simple differential equation. Simply find an anti-derivative F of f, that is a function F such that $F' = f$, and then your solution is

$$y = F(x).$$

But suppose F cannot be given in closed form. That is, it cannot be expressed in terms of sines, cosines, exponentials, logarithms, and other functions for which you have keys on your calculator. In short, F cannot be keyed in on your calculator. This situation is discussed in Chap. X. So, here you are with the very simplest type of differential equation possible, but there is no means to calculate the solution on your calculator.

Don't think the same thing cannot happen with more complicated differential equations. Look at the Example in Sect. 18-4 of T-F. There is posed a problem, and as usual a solution is supplied. Let us interchange x and y in that problem. It becomes

$$(x^2 + y^2)dy + 2xy\,dx = 0 .$$

Can we get a solution of this? Certainly. Just interchange x and y in the solution that T-F got for theirs. If we wish to satisfy the additional condition that $y = 1$ when $x = 1$, the solution is given by

$$y(y^2 + 3x^2) = 4 .$$

Does this give y in closed form? Not at all. To find what y is for a given x one must find a root of a third degree polynomial equation. No calculator has a key for that. Of course, we could review Chap. VII and write a program to get a root by Newton's method. But this is not a triviality. If values of y for several values of x are needed, the methods of this chapter would be more efficient.

But there can be cases considerably worse than this. It is for such cases that the methods of this chapter were devised.

This is a common situation for the differential equations that arise in engineering and the quantitative sciences. If it is your intention to specialize in one of these areas, it will be very helpful to know the material of this chapter. A good time to learn it is when you are studying differential equations in your calculus course, and have such matters fresh in your mind.

1. The Euler method.

We start by looking at first order differential equations. Typical ones are

$$(x^2 + y^2)dy + 2xy\,dx = 0 , \tag{1.1}$$

$$x\frac{dy}{dx} - 3y = x^2 , \tag{1.2}$$

$$x\,dy - y\,dx = xy^2\,dx ; \tag{1.3}$$

these are taken from T-F, namely (1.1) is in the Example in Sect. 18-4 (but with x and y interchanged), (1.2) is in Example 2 in Sect. 18-5, and (1.3) is in the Example in Sect. 18-6. The solutions are

(1.4) $$y(y^2 + 3x^2) = C,$$

(1.5) $$y = -x^2 + Cx^3,$$

(1.6) $$\frac{x}{y} + \frac{x^2}{2} = C,$$

respectively.

Each of these solutions involves an arbitrary constant, C. In order to determine it, it is customary to specify for each first order differential equation a value for y to go with some specified value of x; these are usually called initial conditions or boundary conditions. Thus, let us set the initial conditions

(1.7) $$y = 1 \quad \text{when} \quad x = 1,$$

(1.8) $$y = 3 \quad \text{when} \quad x = 1,$$

(1.9) $$y = 1 \quad \text{when} \quad x = 2,$$

respectively. This will make $C = 4$ in all three cases.

Each first order differential equation can be put in the form

(1.10) $$\frac{dy}{dx} = f(x, y).$$

For the equations (1.1), (1.2), and (1.3) the f's are given by

(1.11) $$f(x,y) = -\frac{2xy}{x^2 + y^2},$$

(1.12) $$f(x, y) = \frac{1}{x}(3y + x^2),$$

(1.13) $$f(x, y) = \frac{1}{x}(y + xy^2),$$

respectively. So the problem we are undertaking is, for some specified f, to find a solution of (1.10), subject to some initial condition

(1.14) $$y = y_0 \quad \text{when} \quad x = x_0.$$

Although it is not proved in most calculus texts, it is the case that, under certain assumptions on f, (1.10) and (1.14) determine a function g such that

(1.15) $$y = g(x)$$

is a solution of (1. 10) and (1. 14). If one forms dy/dx from (1. 15) one will get the same thing that one will get by putting $g(x)$ for y in $f(x, y)$. And, of course $y_0 = g(x_0)$.

If (1. 10) is given in a calculus text, the author has usually been careful to choose it so that g can be given in closed form. But in practical problems, g often cannot be given in closed form. That is the problem with which we must cope.

Sometimes all that is required is an approximation for $g(x_N)$, or for several such. Sometimes it is desired to find out quite a bit about g, and a table of g is called for. At least, there is no trouble about getting the derivative of g. It is given by (1. 10). We will tell how to construct a table. If all that is required is values of y for a few selected values of x, one just arranges that these selected values will appear in the table.

Say we wish a table from x_0 out to x_N. We already know the first entry in the table: y_0 goes with x_0. We choose x_r's with $x_0 < x_1 < x_2 < \ldots < x_{N-1} < x_N$. We refer to the corresponding y's as $y_0, y_1, y_2, \ldots, y_{N-1}, y_N$. That is

(1. 16) $$y_r = g(x_r).$$

We set

(1. 17) $$h_r = x_{r+1} - x_r.$$

We call h_r the step size. Commonly, one takes a lot of the h_r's equal to each other. This corresponds to equal spacing of the x_r's. Tables are customarily made up this way. So we often write just h instead of h_r.

To get to y_1, we proceed as follows. Take

(1. 18) $$\Delta y = y_1 - y_0, \qquad \Delta x = x_1 - x_0.$$

Using Formula 1. 5 from Chap. VI, we have

(1. 19) $$y_0 + \Delta y \cong y_0 + g'(x_0)\Delta x.$$

As we said above, there is no problem about getting $g'(x_0)$. By (1. 10), it is $f(x_0, y_0)$. So we combine (1. 17), (1. 18) and (1. 19) to get

(1. 20) $$y_1 \cong y_0 + h_0 f(x_0, y_0).$$

But now we know y_1, at least approximately. So we can just start over again, getting

(1. 21) $$y_2 \cong y_1 + h_1 f(x_1, y_1).$$

Then, of course,

(1. 22) $$y_3 \cong y_2 + h_2 f(x_2, y_2) .$$

and in general

(1. 23) $$y_{r+1} \cong y_r + h_r f(x_r, y_r) .$$

If we take the h_r's all equal, as is customary, it is no problem to write a program to carry out the calculation indicated in (1. 23). Start with x_r, h, and y_r stored at convenient places. Make up a program that will calculate $f(x_r, y_r)$ from the stored values of x_r and y_r. Then multiply by h, and add to y_r. This gives y_{r+1}. Store it in place of y_r. Add h to x_r. This gives x_{r+1}. Store it in place of x_r. Now you are ready to start over again.

If you wish to change the h_r's, it is a little more complicated, but not much. You calculate $f(x_r, y_r)$ by an appropriate subroutine, and then proceed to y_{r+1} and x_{r+1} as above. Then you calculate h_{r+1} by another appropriate subroutine, and put it in place of h_r.

This is called the <u>Euler method</u>. It is almost ridiculously simple. However, it is not all that accurate, unless you take the h_r's very small. But if you do that you have a lot of steps, and hence a long calculation, before you get to x_N. The formula (1. 19) is not highly accurate unless Δx is quite small. That inaccuracy is built into (1. 20), (1. 21), (1. 22), and every use of (1. 23). So y_1 is not exact. In (1. 21), another error is compounded with that of y_1. And so on.

In Thm. 6. 1 of Sect. 6. 4 of "Elementary Numerical Analysis," second edition, by S. D. Conte and Carl de Boor, McGraw-Hill Book Co. , 1972, it is shown that if one uses a constant h from x_0 to x_N and if $f(x, y)$ and its partial derivatives with respect to x and y are bounded from $x = x_0$ to $x = x_N$, then the error at $x = x_N$ by Euler's method is of the order of h. That is, suppose we go through with one value of h, and then go through again with h half as much. The error the second time should be about half that of the first time.

Let us try the Euler method for the equation (1. 1), with the initial condition $y = 1$ when $x = 1$. Let us try to find out approximately what y is when $x = 2$. We know the solution, in semi closed form, namely (1. 4) with $C = 4$. So all we have to do is find a root of the equation

(1. 24) $$y^2 + 12y - 4 = 0 .$$

There is only one root, and it is approximately

(1. 25) $$0.33032\ 96009 .$$

(See Prob. 4. 10 of Chap. VII.)

Now let us try the Euler method. Our $f(x, y)$ appears on the right of (1. 11). We start at $x = 1$, $y = 1$. Let us try various h's, to see how they work. Start with $h = 1$.

Then $x_1 = 2$, and y_1 should approximate the final answer. Putting $x_0 = y_0 = h_1 = 1$ in (1. 20) gives $y_1 = 0$. This is not very close to (1. 25), but surely you didn't expect to get a very good answer with such a large h.

Now try $h = 0.5$. Then $x_1 = 1.5$ and $x_2 = 2$. So y_2 should approximate the final answer. By (1. 20), $y_1 = 0.5$ and by (1. 21), $y_2 = 0.2$. As predicted, the error has been cut about in half (slightly better, in fact). In Table 1. 1 we have listed the values of y_N, and its errors, for $h = 1, 0.5, 0.25$, and 0.125. It is clear that the errors are

Table 1. 1

h	y_N	error
1	0	0. 33033
0. 5	0. 2	0. 13033
0. 25	0. 27066	0. 05967
0. 125	0. 30149	0. 02884

successively about half as large each time. Incidentally, if we had recorded the y_r's for $h = 0.125$, we would have an approximate table of y.

Problem 1. 1. Solve (1. 2) by the Euler method subject to the initial condition $y = 3$ when $x = 1$, from $x = 1$ out to $x = 2$. Make a table like Table 1. 1 giving y_N and the error for each of $h = 1, 0.5, 0.25$, and 0.125.

Problem 1. 2. Solve (1. 3) by the Euler method subject to the initial condition $y = 1$ when $x = 2$, from $x = 2$ out to $x = 3$. Make a table like Table 1. 1 giving y_N and the error for each of $h = 1, 0.5, 0.25$, and 0.125.

Remark. Clearly something is badly wrong. The y_N values in your table confirm this, without the necessity of looking at the errors.

To see what is happening, recall that the solution is given by (1. 6), with $C = 4$. From this we see that y goes to infinity as x approaches $\sqrt{8} \cong 2.828$. Note that the derivative of y is given by the right side of (1. 13). Because this contains y^2, if y gets large then its derivative gets very large, hastening the ascent to infinity.

If one tries to use the Euler method out to $x = 3$ for (1. 3), one will get some numbers, as you did, but they mean nothing. The theorem we cited about the error being of the order of h depended on $f(x, y)$ (and its derivatives) being bounded. But at $x = \sqrt{8}$, $f(x, y)$ goes to infinity. The numbers you got by the Euler method were hopelessly in error as you approached and passed $x = \sqrt{8}$.

Problem 1. 3. Solve (1. 3) by Euler's method subject to the initial condition $y = 1$ when $x = 2$, from $x = 2$ to $x = 1$. Make a table like Table 1. 1 giving y_N and the error for each of $h = -1, -0.5, -0.25$, and -0.125.

Actually, the solution of (1. 3) for $x > 2$ is not all that intractable. Once we know that y is going to infinity, we set

$$u = \frac{1}{y} . \tag{1.26}$$

Then u will go to zero. The differential equation (1. 3) becomes

$$-x\,du - u\,dx = x\,dx \tag{1.27}$$

or

$$\frac{du}{dx} = -\frac{u+x}{x} . \tag{1.28}$$

The initial condition becomes $u = 1$ when $x = 2$. With this, which gives $C = 4$ in (1. 6), we get from (1. 6) a solution in closed form for u, namely

$$u = \frac{8-x^2}{x} . \tag{1.29}$$

After this transformation, solution by the Euler method proceeds very smoothly.

Problem 1. 4. Solve (1. 28) by Euler's method subject to the initial condition $u = 1$ when $x = 2$, from $x = 2$ to $x = 3$. Make a table like Table 1. 1 giving u_N and the error for each of $h = 1$, 0. 5, 0. 25, and 0. 125.

Moral. When you run into trouble, as in Prob. 1. 2, stop and try some ideas from calculus.

2. The improved Euler method.

Though Table 1. 1 shows that the Euler method is behaving nicely, it is clear that if one wishes to get a quite accurate answer, one would have to take h so small that one would get into one of those all night calculations. So we will devise an improved Euler method.

Obviously we can concentrate on getting an improved estimate for y_1. If we succeed, we then do the same for the rest. The formula (1. 20) gave an estimate of y_1. We will stop and improve it before we go on to y_2. Temporarily call the estimate from (1. 20) $\overline{y}_1$.

By (1. 15)

$$y_1 = y_0 + \int_{x_0}^{x_1} g'(x)\,dx . \tag{2.1}$$

Let us approximate the integral by the trapezoidal rule, taking $N = 1$ in Formula (3. 5) of Chap. X. This gives

(2. 2) $$y_1 \cong y_0 + \frac{h_0}{2}\{g'(x_0) + g'(x_1)\}.$$

But recall that $g'(x)$ is given by $f(x, y)$. So

(2. 3) $$y_1 \cong y_0 + \frac{h_0}{2}\{f(x_0, y_0) + f(x_1, y_1)\}.$$

The difficulty is that y_1 occurs on the right side of this, and we don't yet know y_1. But we know an approximation $\overline{y}_1$ for y_1. Put it in on the right side, giving

(2. 4) $$y_1 \cong y_0 + \frac{h_0}{2}\{f(x_0, y_0) + f(x_1, \overline{y}_1)\}.$$

Using the value of $\overline{y}_1$ from (1. 20) gives finally

(2. 5) $$y_1 \cong y_0 + \frac{h_0}{2}\{f(x_0, y_0) + f(x_0 + h_0, y_0 + h_0 f(x_0, y_0))\}.$$

As we said, we use this idea at every step, so we have

(2. 6) $$y_{r+1} \cong y_r + \frac{h_r}{2}\{f(x_r, y_r) + f(x_r + h_r, y_r + h_r f(x_r, y_r))\}.$$

This is called the improved Euler method, or the modified Euler method, or the Heun method.

For calculations, it is convenient to write this in the form

(2. 7) $$y_{r+1} = y_r + \frac{1}{2}(k_1 + k_2)$$

where

(2. 8) $$k_1 = h_r f(x_r, y_r)$$

(2. 9) $$k_2 = h_r (x_r + h_r, y_r + k_1).$$

For specific functions, f, there are undoubtedly all kinds of short cuts possible in this calculation. But, for a general function, f, one would expect to calculate $f(x,y)$ in some subroutine or extra calculation, using x and y as stored in some fixed memory registers, say in registers one and two. That is, for calculating k_1, x_r and y_r would be read from registers one and two, respectively, and for calculating k_2, $x_r + h_r$ and $y_r + k_1$ would be read from registers one and two, respectively. In addition, h_r is stored in register zero.

The typical program then proceeds as follows:

2. Improved Euler method

RPN

(2.10) [RCL 2], 2, [×], [STO 3],

calculate f(x, y) using the stored x, y, and put into display,

[RCL 0], [STO + 1], [×], [STO + 2], [STO + 3],

calculate f(x, y), using the stored x, y, and put into display,

[RCL 0], [×], [STO + 3], [RCL 3], 2, [÷], [STO 2], [R/S].

RPN

AE

(2.10) [RCL 2], [×], 2, [=], [STO 3],

calculate f(x, y), using the stored x, y, and put into display,

[×], [RCL 0], [STO + 1], [=], [STO + 2], [STO + 3],

calculate f(x, y), using the stored x, y, and put into display,

[×], [RCL 0], [=], [STO + 3], [RCL 3], [÷], 2, [=], [STO 2], [R/S].

AE

Instructions like [STO + 1] refer to register arithmetic. See Sect. 2 of Chap. 0 for details.

The second and fourth lines in Program (2.10) will customarily be done by calling a subroutine which has been stored some place. We have not been more specific because calculators vary so much in their procedures for calling subroutines.

If one wishes to change h_r for the next use of (2.10), one should arrange a suitable means to get h_{r+1} into register zero at the end of (2.10). Usually h_r stays fixed, $h_r = h$, and this detail need not be taken care of.

It is rather difficult to assess the error of the improved Euler method. In Sect. 6.5 of the reference cited in Sect. 1 above, it is stated that if f(x, y) and related functions are well behaved enough, then the improved Euler method is of the order of h^2. In other words, halving h should cut the error by about a factor of 4. This is borne out in Table 2.1, where we have recorded the results of the improved Euler method for the

Table 2.1

h	Y_{1N}	error
1	0.5	-0.16067
0.5	0.36481	-0.03449
0.25	0.33770	-0.00737
0.125	0.33201	-0.00168

same problem,

$$y' = -\frac{2xy}{x^2 + y^2}$$

$$y(1) = 1$$

$$y_N = y(2) ,$$

that we did earlier by the Euler method; see Table 1. 1 for a record of these earlier results.

Those students with nonprogrammable calculators should do only those cases where $|h| \geq 0.25$ in the next three problems.

Problem 2. 1. Do Prob. 1. 1 again, only with the improved Euler method.

Problem 2. 2. Do Prob. 1. 3 again, only with the improved Euler method.

Problem 2. 3. Do Prob. 1. 4 again, only with the improved Euler method.

Consider a water tower that consists of a sphere of radius 20 feet on top of a standpipe of height 40 feet; see Fig. 2. 1. At the bottom of the standpipe is a valve that can be opened to produce a circular orifice of diameter 6 inches. If there is water in the tower three fourths the way up in the sphere, and we open the orifice, how long will it take for enough water to drain out that the water is only halfway up in the sphere?

That is, if H is the distance of the top of the water from the ground, and we start with $H = 70$, how long will it take to get the level down to $H = 60$?

Torricelli's Theorem states that the flow rate through the orifice is

$$\alpha a\sqrt{2gH} , \tag{2.11}$$

where α is a contraction coefficient, a is the area of the orifice, g is the acceleration due to gravity (take it to be 32. 16 feet per $\sec^2$), and H is the vertical distance from the orifice up to the top surface of the water. Commonly α is in the neighborhood of 0. 5. If we denote the total volume of water in the tower by V, we have

$$\frac{dV}{dt} = -(0.5)a\sqrt{(64.32)H} . \tag{2.12}$$

If we take a horizontal slice of the sphere dH thick (see Fig. 2. 1), we will have approximately

$$dV \cong \pi((20)^2 - (H - 60)^2)dH .$$

Taking the limit as dH goes to zero gives

2. Improved Euler method

Figure 2. 1

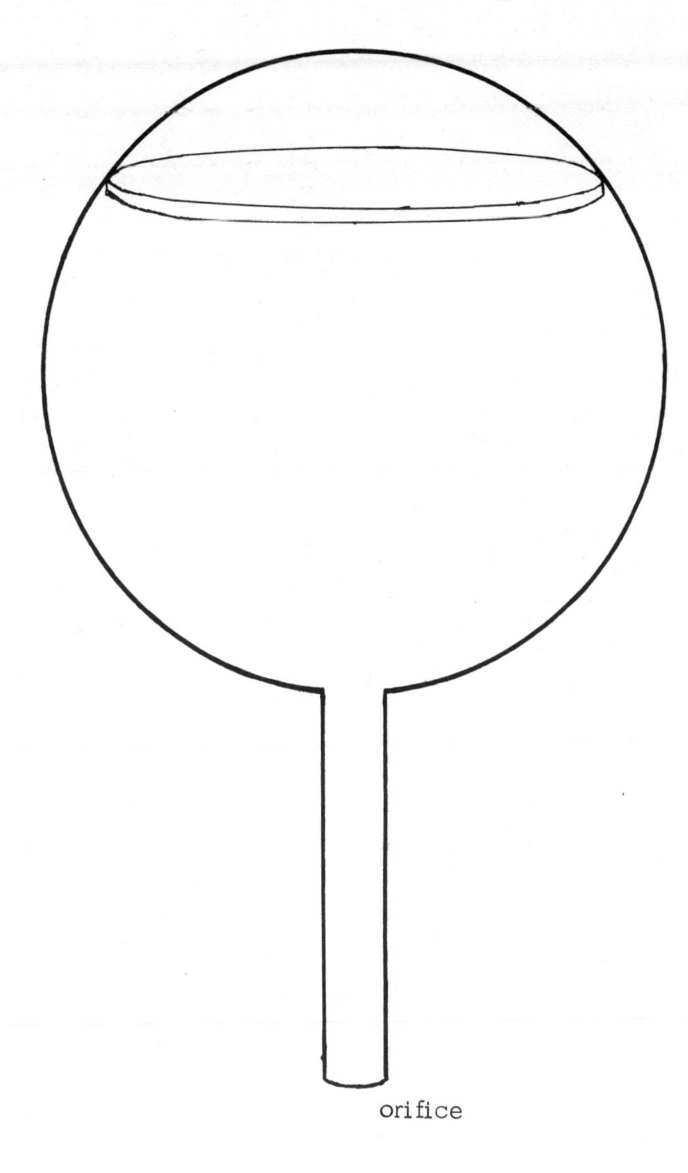

(2. 13) $$\frac{dV}{dH} = \pi(400 - (H - 60)^2) .$$

Dividing this into (2. 12) gives

(2. 14) $$\frac{dH}{dt} = -\frac{(0.5)a\sqrt{(64.32)H}}{\pi(400 - (H - 60)^2)} .$$

This is a differential equation for H in terms of t. By taking step sizes of $h = \Delta t$, we can use the improved Euler method to make a table of H as a function of t. Then we look to see for what value of t we get $H = 60$. Of course, this is not going to come out exactly at any entry in the table. But, when we get a value of $H > 60$ for one t and a value of $H < 60$ for the next t, we can interpolate, by the methods of Sect. 3 of Chap. VI, to find a value of t for which approximately $H = 60$.

The next problem should not be attempted by anyone who does not have a programmable calculator.

*Problem 2. 4. For the situation described above, find out approximately how long it will take for H to get from 70 feet down to 60 feet with the orifice open.

Practical problems have a way of being rather messy. To get a reasonably accurate value of the time, you will need to use a fair number of steps. Actually, your calculator has features that make this fairly easy. You will be using Program (2. 10), or an equivalent. Put in two other features: (a) arrange to store not only the H for the current value of t, but also the previous one; (b) fix it so that the calculator will test if $H < 60$ after each step - if it is not, it goes on to the next step automatically, but if it is, it stops. Set up the calculation with a fairly small step size, and leave the calculator running. It will stop the first time it gets to an $H < 60$. Then you interpolate between that and the previous one.

Don't break your back about trying to get a really accurate value of t. By (2. 11), the rate of flow depends on the contraction coefficient, α. This is seldom known very accurately. The value of 0. 5 that we took for α in (2. 12) could easily be off by several per cent. So, even if we could get an exact solution of (2. 14), this exact solution could differ by several per cent from the actual time from $H = 70$ to $H = 60$. So, get a reasonably accurate value of t, and relax.

So your calculator can make a fairly easy problem out of this. But what about your brain, and all this calculus that you have been learning? You are supposed to stay alert and watch out for ways to make a problem really easy. Like the following.

Turn the equation (2. 14) upside down, getting

(2. 15) $$\frac{dt}{dH} = -\frac{\pi(400 - (H - 60)^2)}{(0.5)a\sqrt{(64.32)H}} .$$

Now you have a differential equation for t as a function of H. But notice that t does not appear on the right side of (2. 15). All you have to do to get t is to integrate. The time from H = 70 to H = 60 is just

$$\int_{70}^{60} \left\{ -\frac{\pi(400 - (H-60)^2)}{(0.5)a\sqrt{(64.32)H}} \right\} dH ,$$

which is the same as

$$\int_{60}^{70} \frac{\pi(400 - (H-60)^2)}{(0.5)a\sqrt{(64.32)H}} \, dH . \tag{2.16}$$

You can approximate this easily by the methods of Chap. X, on numerical integration.

Even the reader with a nonprogrammable calculator should find the next problem very easy.

Problem 2. 5. Approximate (2. 16) by the simplest Simpson rule, namely (2. 7) of Chap. X.

Moral. If you don't have a fancy calculator, make more use of ideas from calculus.

The results in Table 2. 1 are a great improvement over those in Table 1. 1. This was achieved at a price. One has to evaluate the function $f(x, y)$ once per step in the Euler method. For $h = 0.125$ in Table 1. 1, eight function evaluations were required. By (2. 6) one must evaluate $f(x, y)$ two times per step in the improved Euler method. For $h = 0.125$ in Table 2. 1, sixteen function evaluations were required.

Let us look again at the derivation of (2. 5). We got this by putting a guess for y_1, from (1. 20), on the right side of (2. 3). But this gives us a better guess for y_1. Suppose we put this better guess on the right side of (2. 3). It would cost us another function evaluation, since we must now calculate $f(x_1, y_1)$ for the improved y_1. But it does give a still better guess for y_1, but not a lot better. We would now be up to 24 function evaluations for the problem of Tables 1. 1 and 2. 1 for $h = 0.125$. Can we do better with 24 function evaluations? Indeed we can.

We could take $h = 1/12$ and make use of (2. 6) twelve times. That would cut the overall error by more than a half.

If we really wish a great improvement, we could go to a Runge-Kutta method of order 4. This would cost us 32 function evaluations to get from $x = 1$ to $x = 2$ with $h = 0.125$.

3. A Runge-Kutta method of order 4.

There are Runge-Kutta methods of all orders, and many methods for any given order. Actually the Euler method is a Runge-Kutta method of order 1 and the improved Euler method is a Runge-Kutta method of order 2. See Sect. 6.5 of the reference cited in Sect. 1 above. There also is given the most popular Runge-Kutta method of order 4, namely:

$$y_{r+1} = y_r + \frac{1}{6}(k_1 + 2k_2 + 2k_3 + k_4) \tag{3.1}$$

where

$$k_1 = hf(x_r, y_r) \tag{3.2}$$

$$k_2 = hf(x_r + \frac{h}{2}, y_r + \frac{1}{2}k_1) \tag{3.3}$$

$$k_3 = hf(x_r + \frac{h}{2}, y_r + \frac{1}{2}k_2) \tag{3.4}$$

$$k_4 = hf(x_r + h, y_r + k_3) . \tag{3.5}$$

We have written h for h_r. It is quite in order to change the step size, h_r, for each step. Obviously, each step requires 4 function evaluations.

As the name "Runge-Kutta method of order 4" would suggest, this is a method of the order of h^4. In other words, halving h should cut the error by about a factor of 16. This is borne out in Table 3.1, where we have recorded the results of the Runge-Kutta

Table 3.1

h	y_N	error
1	0.33945 32740	$-9.1236\ 731 \times 10^{-3}$
0.5	0.33077 53580	$-4.4575\ 71 \times 10^{-4}$
0.25	0.33036 04541	$-3.0853\ 2 \times 10^{-5}$
0.125	0.33033 15223	-1.9214×10^{-6}

method of order 4 for the same problem,

$$y' = -\frac{2xy}{x^2 + y^2}$$

$$y(1) = 1$$

$$y_N = y(2) ,$$

that we did earlier; see Tables 1. 1 and 2. 1.

We rounded the intermediate y_r's for $h = 0.125$ and recorded them in Table 3. 2, which constitutes a short table of the solution of (1. 1), correct to 5 decimals.

Table 3. 2

x	y	x	y
1. 000	1. 00000	1. 625	0. 49008
1. 125	0. 87628	1. 750	0. 42691
1. 250	0. 75977	1. 875	0. 37429
1. 375	0. 65556	2. 000	0. 33033
1. 500	0. 56577		

In programming the Runge-Kutta method of order 4, we start out with x_r and y_r stored, in registers one and two, say. As to $h = h_r$, it is more convenient here to store $h/2$ rather than h, in register zero, say. In addition, we use a register, five say, to store y_r temporarily, and a register, four say, to build up the sum $k_1 + 2k_2 + 2k_3 + k_4$. Note that we intend to multiply by $h/2$ rather than h every time the factor h is called for in (3. 2) to (3. 5), so that instead of calculating k_1, k_2, k_3, and k_4 we calculate their halves. Hence we would in calculating (3. 1) only divide by 3 rather than by 6.

Here is a program.

RPN

(3. 6) [STO 5],

calculate f(x, y), using the stored x, y, and put into display,

[RCL 0], [STO + 1], [×], [STO 4], [STO + 2],

calculate f(x, y), using the stored x, y, and put into display,

[RCL 0], [×], [STO + 4], [STO + 4], [RCL 5], [+], [STO 2],

calculate f(x, y), using the stored x, y, and put into display,

[RCL 0], [STO + 1], [×], 2, [×], [STO + 4], [RCL 5], [+], [STO 2],

calculate f(x, y), using the stored x, y, and put into display,

[RCL 0], [×], [RCL 4], [+], 3, [÷], [RCL 5], [+], [STO 2], [R/S].

RPN

AE

(3. 6) [STO 5],

calculate f(x, y), using the stored x, y, and put into display,

[×], [RCL 0], [STO + 1], [=], [STO 4], [STO + 2],

calculate f(x, y), using the stored x, y, and put into display,

[×], [RCL 0], [=], [STO + 4], [STO + 4], [+], [RCL 5], [=],

[STO 2],

calculate f(x, y), using the stored x, y, and put into display,

[×], [RCL 0], [STO + 1], [×], 2, [=], [STO + 4], [+],

[RCL 5], [=], [STO 2],

calculate f(x, y), using the stored x, y, and put into display,

[×], [RCL 0], [+], [RCL 4], [=], [÷], 3, [+], [RCL 5], [=],

[STO 2], [R/S].

AE

After execution of Program (3. 6), you will have x_{r+1} and y_{r+1} in registers one and two, respectively, in place of x_r and y_r. You will have y_{r+1} also in the display, ready for the first step of Program (3. 6), which stores what is in the display into register five.

CAUTION. At the very beginning, when Program (3. 6) has not yet been run, you have to be sure to get y_0 into the display. But this is easily done. Input $h/2$ and store it into register zero, then input x_0 and store it into register one, and finally input y_0 and store it into register two. Then there is y_0 sitting in the display, ready for you to run Program (3. 6).

You will put the calculation of f(x, y), from values of x and y stored in registers one and two, respectively, into a subroutine, to be called four times. While you are at it, you might as well put a little more into the subroutine.

RPN

In the RPN version of Program (3. 6), tack an extra step [RCL 0] onto the end of the subroutine. Then the subroutine will take care of four [RCL 0]'s that appear in Program (3. 6) as written above, at the cost of one [RCL 0] in the subroutine. So we have saved three program steps in the program. Altogether, the bookkeeping details of Program (3. 6), including the [RCL 0] in the subroutine use up 33 program steps (including 4 subroutine calls not shown), leaving adequate room to program the calculation of f(x, y).

RPN

AE

In the AE version of Program (3. 6), tack two extra steps, [×], [RCL 0] onto the end of the subroutine. Then the subroutine will take care of four pairs, [×], [RCL 0], that appear in Program (3. 6) as written above, at the cost of a pair, [×], [RCL 0] in the subroutine. So we have saved six program steps in the program. Altogether, the bookkeeping details of Program (3. 6), including the [×], [RCL 0] in the subroutine use up 37 program steps (including 4 subroutine calls not shown), leaving a good deal of room to program the calculation of $f(x,y)$.

AE

If h_{r+1} is different from h_r, a change from having $h_r/2$ stored in register zero to having $h_{r+1}/2$ stored there can be done at the end of Program (3. 6).

Those students with nonprogrammable calculators should do only those cases where $|h| \geq 0.5$ in the next three problems.

Problem 3. 1. Do Prob. 1. 1 again, only with the Runge-Kutta method of order 4.

Hint. A fairly economical program for calculating the right side of (1. 12) and putting it into the display is:

3, [STO 3], [RCL 2], [STO × 3], [RCL 1],

[STO ÷ 3], [STO + 3], [RCL 3].

Problem 3. 2. Do Prob. 1. 3 again, only with the Runge-Kutta method of order 4.

Problem 3. 3. Do Prob. 1. 4 again, only with the Runge-Kutta method of order 4.

With the methods you have learned, you can handle a majority of the first order differential equations you are liable to encounter. However, you have not learned all the methods there are, even by half. If you have trouble with a differential equation, as earlier with Prob. 1. 2, try a transformation, or some trick from calculus. If this fails, consult a professional numerical analyst. There are some differential equations of such difficulty as to tax a very large, very fast, very expensive computer, even using quite sophisticated methods. If you have happened onto one of these, you need help. However, first try all the calculus tricks you can think of.

4. Second order equations.

These are discussed in Sections 18-7, 18-9, and 18-10 of T-F. They can usually be written in the form

$$\frac{d^2y}{dx^2} = f(x, y, \frac{dy}{dx}) . \tag{4. 1}$$

For an initial condition, one usually specifies a value for both y and dy/dx for some specified value of x.

Thus, consider the first two problems at the end of Sect. 18-10 of T-F, namely:

(4. 2) $$\frac{d^2y}{dx^2} + \frac{dy}{dx} = x$$

(4. 3) $$\frac{d^2y}{dx^2} + y = \tan x \ .$$

For them, $f(x, y, \frac{dy}{dx})$ is

(4. 4) $$f(x, y, \frac{dy}{dx}) = x - \frac{dy}{dx}$$

(4. 5) $$f(x, y, \frac{dy}{dx}) = \tan x - y$$

respectively. Let us set the initial conditions

(4. 6) $$y = \frac{1}{2}, \ \frac{dy}{dx} = -1 \quad \text{when} \quad x = 0$$

(4. 7) $$y = 0, \ \frac{dy}{dx} = -1 \quad \text{when} \quad x = 0$$

respectively. Then the solutions are

(4. 8) $$y = \frac{1}{2}(x - 1)^2$$

(4. 9) $$y = - \cos x \sinh^{-1}(\tan x)$$

respectively.

As we did for first order equations, we choose values of x, namely $x_0, x_1, x_2, \ldots, x_{N-1}, x_N$, where x_0 is the x for the initial condition. Corresponding to them, we try to find approximations for y and dy/dx, namely $y_0, y_1, y_2, \ldots, y_{N-1}, y_N$ and $y'_0, y'_1, y'_2, \ldots, y'_{N-1}, y'_N$.

Analogously to the Euler method, we could use the approximations

(4. 10) $$y'_{r+1} \cong y'_r + h_r f(x_r, y_r, y'_r)$$

(4. 11) $$y_{r+1} \cong y_r + h_r y'_r \ .$$

For the equation (4. 2) we have $y = 0 = y'$ at $x = 1$. If we use (4. 10) and (4. 11), we get the results shown in Table 4. 1. This seems to confirm that the errors are of the order of h.

Let us try for an analog of the improved Euler method. For this, we use the two

Table 4. 1

h	y_N	error	y'_N	error
1	-0. 5	0. 5	0	0
0. 5	-0. 25	0. 25	0	0
0. 25	-0. 125	0. 125	0	0
0. 125	-0. 0625	0. 0625	0	0

trapezoidal rule approximations

$$y'_{r+1} \cong y'_r + \frac{h_r}{2}\{f(x_r, y_r, y'_r) + f(x_{r+1}, y_{r+1}, y'_{r+1})\} \tag{4. 12}$$

$$y_{r+1} \cong y_r + \frac{h_r}{2}\{y'_r + y'_{r+1}\} . \tag{4. 13}$$

The difficulty, as before, is that we don't have the values for y'_{r+1} and y_{r+1} to put in the right sides of these. So we use some approximations instead, namely

$$\hat{y}'_{r+1} = y'_r + f_r ,$$

$$\hat{y}_{r+1} = y_r + \frac{h_r}{2}\{y'_r + \hat{y}'_{r+1}\} ,$$

with $f_r = h_r f(x_r, y_r, y'_r)$.

Here, $\hat{y}'_{r+1}$ is the approximation for y'_{r+1} obtained from (4. 10), and $\hat{y}_{r+1}$ is the result of substituting $\hat{y}'_{r+1}$ for y'_{r+1} in (4. 13). With these, (4. 12) gives the still better approximation

$$y'_{r+1} \cong y'_r + \frac{1}{2}\{f_r + h_r f(x_{r+1}, \hat{y}_{r+1}, \hat{y}'_{r+1})\} . \tag{4. 14}$$

Finally, we use this value of y'_{r+1} in (4. 13) to get y_{r+1}.

Using this procedure for (4. 2), we get the results shown in Table 4. 2. This seems to confirm that the errors are of the order of h^2.

Table 4. 2

h	y_N	error	y'_N	error
1	0	0	0	0
0. 5	0	0	0	0
0. 25	0	0	0	0
0. 125	0	0	0	0

If Table 4.2 looks too good to be true, recall that by (4.8) the solution of (4.2) is a polynomial of degree two. So a method of order two could reasonably hit the answers right on the nose, which is exactly what happened. By (4.8), $y' = x - 1$. This is a first degree polynomial, so that a method of order one could hit y' right on the nose. By Table 4.1, that is just what happened when we used (4.10) and (4.11).

Suppose we solve (4.3) with the initial conditions (4.7). What should y and y' be when $x = 1$? In (4.9) we have a formula for y. The function $\sinh^{-1}$ is not given on all calculators. However, it is not all that recondite. By (10) in Sect. 9-5 of T-F,

(4.15) $$\sinh^{-1} x = \ln (x + \sqrt{x^2 + 1}) .$$

So

(4.16) $$\sinh^{-1} (\tan x) = \ln \left(\frac{1+\sin x}{\cos x}\right) .$$

So for $x = 1$, we have

(4.17 $$y \cong -0.66251\ 39172$$

(4.18) $$y' \cong 0.03180\ 42920 .$$

If you have only a nonprogrammable calculator, omit the case $h = 0.125$ in the next problem.

Problem 4.1. Solve (4.3) by means of (4.13) and (4.14) subject to the initial condition $y = 0$ and $y' = -1$ when $x = 0$, from $x = 0$ out to $x = 1$. Make a table like Table 4.2 giving y_N and y'_N and their errors for each of $h = 1, 0.5, 0.25$, and 0.125.

In Sect. 18-12, T-F have quite a discussion of vibrations, confining their attention almost exclusively to equations for which the solution can be given in closed form. However, in Prob. 4 at the end of the section, they take up the case of a pendulum, for which the differential equation is

(4.19) $$\frac{d^2\theta}{dt^2} = -\frac{g}{\ell} \sin \theta ;$$

here ℓ is the length of the pendulum, g is the acceleration due to gravity, and θ is the angle of the pendulum, measured counterclockwise from the vertical line drawn down from the point of support of the pendulum. Let us consider the case where $\ell = g$, reducing (4.19) to

(4.20) $$\frac{d^2\theta}{dt^2} = - \sin \theta .$$

Of course, T-F immediately simplify this by assuming that θ remains so small that we may approximate $\sin \theta$ by θ. In this case, if the pendulum reaches its rightmost swing, of amount θ_0, at time $t = 0$, the solution of the simplified equation is

(4. 21) $$\theta = \theta_0 \cos t \; .$$

Let us consider the following problem. Swing the pendulum to the right, clear up to where it is horizontal $(\theta = \pi/2)$. At time $t = 0$, let it go. How long will it take to swing down to a vertical position $(\theta = 0)$?

We have taken θ so large that $\sin\theta$ could not be replaced by θ. So the simplified solution (4. 21) is not applicable. If it were, θ_0 would be $\pi/2$, and it would say that $\theta = 0$ at $t = \pi/2$.

Problem 4. 2. Solve the problem posed above. That is, solve (4. 20) subject to the initial conditions that $\theta = \pi/2$ and $d\theta/dt = 0$ at $t = 0$. Find out approximately when $\theta = 0$ by interpolating in your solution. Take t at step sizes 0. 2 out to the first point where θ is negative. Record the values of θ and $d\theta/dt$ at each step, and see how they compare with what would be given by the approximate solution (4. 21) with $\theta_0 = \pi/2$.

If you have a nonprogrammable calculator, take your step size to be 0. 4.

If you wish more accuracy, just take a smaller step size. It is not necessary to record the value of θ after each step. You can program the calculator to take a step size of 0. 02, run through 10 steps automatically, and then stop for you to record θ. You would also set it to stop the first time θ is negative, so that you can interpolate. As this is a second order method, this should give about two more decimals of accuracy than you got in Prob. 4. 2.

But hold! Are we not supposed to be alert for easier ways to do problems? Can calculus help us here? Indeed it can. In Sect. 18-7 of T-F, it is suggested that we define

(4. 22) $$p = \frac{d\theta}{dt} ,$$

which will give

(4. 23) $$\frac{d^2\theta}{dt^2} = p\frac{dp}{d\theta} \; .$$

So (4. 20) reduces to

$$p\frac{dp}{d\theta} = - \sin\theta \; .$$

If we integrate both sides with respect to θ, and use the initial condition that $p = 0$ when $\theta = \pi/2$, we will get

$$p^2 = 2\cos\theta \; .$$

So

(4. 24) $$\frac{d\theta}{dt} = -\sqrt{2\cos\theta}\ ;$$

we take the negative square root because θ is decreasing at first.

Turn (4. 24) upside down, and get

$$\frac{dt}{d\theta} = \frac{-1}{\sqrt{2\cos\theta}}\ .$$

So we can find t by an integration. The value of t when $\theta = 0$ is given by

(4. 25) $$\int_{\frac{\pi}{2}}^{0} \frac{-d\theta}{\sqrt{2\cos\theta}} = \int_{0}^{\frac{\pi}{2}} \frac{d\theta}{\sqrt{2\cos\theta}}\ .$$

This is one of the better known examples of a definite integral that cannot be given in closed form. But Chap. X is supposed to tell us how to get a numerical approximation anyhow. However, we encounter a difficulty. The methods of Chap. X depend on the derivatives of the integrand being well behaved. But for (4. 25), the integrand and all its derivatives go to infinity at $\theta = \pi/2$.

In Sect. 7-9 of T- F, it is suggested that one might try various kinds of substitutions. Try putting

(4. 26) $$\theta = \frac{\pi}{2} - \phi^2\ .$$

Then (4. 25) transforms to

(4. 27) $$\sqrt{2}\int_{0}^{\sqrt{\pi/2}} \frac{\phi\, d\phi}{\sqrt{\sin\phi^2}}\ .$$

The integrand of (4. 25) went to infinity for $\theta = \pi/2$, which corresponds to $\phi = 0$. What happens to the integrand of (4. 27) at $\phi = 0$? We have

$$\lim_{\phi\to 0} \frac{\phi}{\sqrt{\sin\phi^2}} = \lim_{\phi\to 0} \sqrt{\frac{\phi^2}{\sin\phi^2}} = \sqrt{1} = 1\ .$$

It can also be shown that the integrand for (4. 27) has continuous derivatives for $0 \le \phi \le \sqrt{\pi/2}$, but this is pretty intricate, so we will skip it.

Anyhow, the methods of Chap. X can be used to get a highly accurate value for (4. 27). Let us settle for something less ambitious.

Problem 4. 3. Approximate (4. 27) by the simplest Simpson rule, namely (2. 7) of Chap. X.

4. Second order equations

In Sect. 18-12 of T-F is considered the case of a vibration where there is friction proportional to the velocity. Let us consider the case where the vibrating object is in a viscuous fluid for which the deceleration caused by the fluid is proportional to the square of the speed. Analogous to the treatment in Sect. 18-12 of T-F, the differential equation is

$$m \frac{d^2x}{dt^2} = -kx \pm c \left(\frac{dx}{dt}\right)^2 ,$$

where c is positive and one takes a minus sign in front of c when dx/dt is positive, and a plus sign when dx/dt is negative. To simplify the present discussion, take $m = k = c = 1$. Then our differential equation is like (4. 1), namely

$$\frac{d^2x}{dt^2} = f\left(t, x, \frac{dx}{dt}\right), \tag{4. 28}$$

with

$$f\left(t, x, \frac{dx}{dt}\right) = \begin{cases} -x - \left(\frac{dx}{dt}\right)^2 & \text{when } \frac{dx}{dt} \geq 0 \\ -x + \left(\frac{dx}{dt}\right)^2 & \text{when } \frac{dx}{dt} \leq 0 . \end{cases} \tag{4. 29}$$

Though this looks rather intractable, it is the case that one can easily program a calculator to calculate f.

<u>Problem 4. 4.</u> Solve the differential equation constituted by (4. 28) and (4. 29), subject to the initial conditions $x = 1$ and $dx/dt = 0$ when $t = 0$. Find approximately the time at which the vibrating object completes its first return to positive x, the time at which $dx/dt = 0$ just before x begins again to decrease.

Hint. Use (4. 13) and (4. 14) with a time step equal to 0. 5, and interpolate to approximate the t for which $dx/dt = 0$.

Can we achieve a significant simplification by appealing to calculus methods? It appears that in this case we cannot. If we wish to go through several vibrations of our object, then dx/dt changes sign several times. So we have to keep changing the sign of

$$\left(\frac{dx}{dt}\right)^2$$

in (4. 29). For calculating f on a calculator, this is no trouble at all. One tests if dx/dt is ≥ 0 or not by a suitable key. If $dx/dt \geq 0$, the calculator goes down one branch of the program. Otherwise, it goes down the other branch. However, if one is trying to handle (4. 29) mathematically, one has to start over again every time dx/dt changes sign. And every time we start over, we have to do a root finding calculation to determine the value of x at which the sign changes, and then a numerical integration to find the value of t. And we wind up with a different formula every time, of course. It is a mess. The procedure in the hint for Prob. 4. 4 can be set up easily. If you wish high accuracy, you take a small time step, arranging for the calculator to run off quite a number of steps, and then stop so that you can keep track of the progress of the solution. And it will then go on to the next stopping place after you press the program key (or keys).

PROGRAM APPENDIX FOR THE HP-33E

Unless specifically stated otherwise, the programs are to be stored as follows. Store the first instruction in program register 01, the next in program register 02, and so on consecutively. Then, to run the program, simply push the R/S key, provided program control is at the beginning of program memory. If it is not, or if you are not sure, press RTN first.

For ease of understanding, we have prefaced each program step to which there is a GTO with the number of the program memory register into which it is supposed to go.

Program II. 1. Evaluation of a polynomial by Horner's method.

This program is first referred to in Sect. 1 of Chap. II. It uses Horner's method to evaluate the polynomial, p, defined by

(II. 1. 1) $$p(x) = a_0x^n + a_1x^{n-1} + \dots + a_{n-1}x + a_n$$

at c. The coefficients $a_0, a_1, \dots, a_n$ are to be input in the course of storing the program. Start with c in the display.

II. 1 [GTO 05],

$_{02}$ [+], [x], [RTN],

$_{05}$ [↑], [↑], [↑], a_0, [×],

a_1, [GSB 02],

a_2, [GSB 02],

. . . .

a_{n-1}, [GSB 02],

a_n, [+], [GTO 00].

At the end of the program, p(c) will be in the display. Note that the last instruction of the program returns control to the beginning, so that if then you wish to calculate $p(c^*)$ for some c^* other than c, just input c^* and press R/S.

Unless $n \geq 6$, this program uses as many steps as, or more steps than, Program (1. 6) in Chap. II, and the latter might better be used.

Program II. 2. Evaluation of a polynomial of seventh degree by Horner's method (with previously stored coefficients).

The coefficients $a_0, a_1, \ldots, a_7$ are to be stored respectively in registers zero, one, ..., seven. Start with c in the display.

II. 2 [GTO 05],

02 [+], [×], [RTN],

05 [↑], [↑], [↑], [RCL 0], [×],

[RCL 1], [GSB 02],

[RCL 2], [GSB 02],

[RCL 3], [GSB 02],

[RCL 4], [GSB 02],

[RCL 5], [GSB 02],

[RCL 6], [GSB 02],

[RCL 7], [+], [GTO 00].

At the end of the program, p(c) will be in the display.

Suppose you wish to evaluate a fifth degree polynomial. You can artifically make a seventh degree polynomial out of it by adding the terms

$$0x^7 + 0x^6 .$$

Then Program II. 2 will evaluate it. Alternatively, you can store $a_0, a_1, \ldots, a_5$ respectively in registers zero, one, ..., five. Then replace the program line

[RCL 5] , [GSB 02]

by

[RCL 5] , [+] , [GTO 00] ,

and omit the final two lines.

If you have a polynomial of degree higher than seven, you can work in a few lines of Program II. 1 toward the end of Program II. 2.

There are occasions when you might wish to see the successive b_r's that are calculated by Horner's method. See Sect. 2 of Chap. II for a discussion of this. To accomplish this, change the first three lines of either Program II. 1 or II. 2 to:

[GTO 06] ,

02 [+] , [R/S] , [×] , [RTN] ,

06 [↑] , [↑] , [↑] , [RCL 0] , [R/S] , [×] .

Now the program will stop every time a b_r appears in the display. After you have done whatever you wish with that b_r, you press R/S, and the program will go on to the next b_r.

Program II. 3. Deflation of a seventh degree polynomial.

Let the coefficients of the polynomial be stored as for Program II. 2. Start with c in the display.

II. 3

[↑] , [↑] , [↑] , [RCL 0] ,

[×] , [RCL 1] , [+] , [STO 1] ,

[×] , [RCL 2] , [+] , [STO 2] ,

[×] , [RCL 3] , [+] , [STO 3] ,

[×] , [RCL 4] , [+] , [STO 4] ,

[×] , [RCL 5] , [+] , [STO 5] ,

[×] , [RCL 6] , [+] , [STO 6] ,

[×] , [RCL 7] , [+] , [GTO 00] .

At the end, b_r has replaced a_r for $r = 0, 1, \ldots, 6$, and $b_7 = p(c)$ is in the display. This program is used when one wishes to deflate a polynomial. See Sect. 2 of Chap. II.

Program II. 4. Evaluation of a seventh degree polynomial and its derivative by Horner's method.

Let the coefficients of the polynomial be stored as for Program II. 2. Start with c in the display.

II. 4 [GTO 13],

02 [+], [R↓], [×], [x⇄y], [R↓], [+], [R↓], [x⇄y], [R↓], [×], [RTN],

13 [↑], [↑], [RCL 0], [x⇄y], [RCL 0], [×],

[RCL 1], [GSB 02],

[RCL 2], [GSB 02],

[RCL 3], [GSB 02],

[RCL 4], [GSB 02],

[RCL 5], [GSB 02],

[RCL 6], [GSB 02],

[RCL 7], [+], [GTO 00].

At the end, p(c) is in the display and p'(c) is in the y position in the stack. If you press x⇄y, then p(c) and p'(c) will change places, bringing p'(c) into the display. Pressing x⇄y again will put them back as they were.

If you have a polynomial of degree less than seven, proceed as explained after Program II. 2.

Program VII. 1. Bisection.

This is a program to carry out the search for a root by the bisection method. See Sect. 3 of Chap. VII. Start with m_r and h_r respectively in registers zero and one. If $f(m_r) = 0$, the program stops with zero in the display. Otherwise, it puts m_{r+1} and h_{r+1} respectively in registers zero and one, and stops with h_{r+1} in the display. As h_{r+1} is never zero, you can tell if you have found a zero, or if you must run the program again.

VII. 1 2, [STO ÷ 1], [GSB 14],

[x = 0], [R/S],

[x > 0], [GTO 1],

[RCL 1], STO -0], [GTO 00],

11 [RCL 1], [STO + 0], [GTO 00],

subroutine .

14

The subroutine depends on what function f you are working with. It should be written to calculate $f(m_r)$, taking m_r from register zero, and to put $f(m_r)$ in the display.

Program VII. 2. Newton's method for polynomial root finding.

This is a program for approximating a zero of a polynomial of degree eight. In the first place, we must have $a_0 \neq 0$, since otherwise the polynomial is of degree seven or less, and we proceed as in Sect. 5 of Chap. VII. So we divide the polynomial by a_0. This does not change the location of any zeros of the polynomial, but it makes the leading coefficient equal to unity. We will take advantage of this.

We first store the coefficients $a_1, a_2, \ldots, a_8$ respectively in registers zero, one, ..., seven. We put a_0 in the display. Then we execute the steps

[STO ÷ 0], [STO ÷ 1], [STO ÷ 2], [STO ÷ 3],

[STO ÷ 4], [STO ÷ 5], [STO ÷ 6], [STO ÷ 7] .

Now, except for the leading coefficient (which is unity), we have stored the coefficients of the polynomial we got by dividing the original polynomial by a_0.

Now we seek a zero by Newton's method (see Sect. 5 of Chap. VII). Let x_r be an approximation to a zero. Put x_r into the display. Next, for the polynomial p that we are now dealing with, we evaluate p and p' both at x_r and stop to see if $p(x_r)$ seems near enough to 0 to satisfy us. If not, we calculate x_{r+1} by (5. 3) of Chap. VII, namely

$$x_{r+1} = x_r - \frac{p(x_r)}{p'(x_r)} ,$$

with x_{r+1} finally in the display, and stop for inspection. But now x_{r+1} is in the display, and we can just repeat the process. After each stop, we simply press R/S to get going again since the program takes care of getting us to the beginning.

VII. 2 [GTO 13],

02 [+], [R↓], [×], [x⇄y], [R↓], [+], [R↓], [x⇄y], [R↓], [×], [RTN],

13 [↑], [↑], [1], [x⇄y],

[RCL 0], [GSB 02],

[RCL 1], [GSB 02],

[RCL 2], [GSB 02],

[RCL 3], [GSB 02],

[RCL 4], [GSB 02],

[RCL 5], [GSB 02],

[RCL 6], [GSB 02],

[RCL 7], [+], [R/S],

[x⇄y], [÷], [−], [GTO 00] .

Program XIII. 1. Damped Newton's method for finding simultaneous solutions of two equations.

The theory for this is given in Sect. 5 of Chap. XIII. We seek points (x, y) such that

$$g(x, y) = 0, \qquad h(x, y) = 0$$

are both true. We have a point (x_r, y_r) which we hope is "close" to a simultaneous solution, and wish to find (x_{r+1}, y_{r+1}) which is "closer." To monitor the closeness, we define

$$E(x, y) = (g(x, y))^2 + (h(x, y))^2 .$$

So we seek (x_{r+1}, y_{r+1}) for which

$$E(x_{r+1}, y_{r+1}) < E(x_r, y_r) .$$

Parts of the program have to be done by pressing keys by hand. Also, we have to keep a written record of $E(x_r, y_r)$, $E(x_{r+1}, y_{r+1}), \ldots$.

(1) We start with x_r, y_r, $-g(x_r, y_r)$, $-h(x_r, y_r)$ respectively stored in registers zero, seven, three, and six, and with $E(x_r, y_r)$ in the display.

(2) Write down $E(x_r, y_r)$.

(3) By hand, calculate $g_x(x_r, y_r)$, $g_y(x_r, y_r)$, $h_x(x_r, y_r)$, $h_y(x_r, y_r)$ and store them respectively in registers one, two, four, and five.

(4) We assume that you have stored Program (3. 7), RPN, of Chap. XIII, with [R/S] added to the end, in program registers 01 through 18.

(5) Press RTN and R/S. This runs Program (3. 7), RPN, of Chap. XIII. This solves the equations (4. 4) of Chap. XIII, and there are now numbers δx and δy respectively in registers three and six. These are proposed corrections for x_r and y_r, so that we should consider taking

$$x_{r+1} = x_r + \delta x, \quad y_{r+1} = y_r + \delta y .$$

But first we have to check if this will decrease E. This requires shuffling $x_r, y_r, \delta x$, and δy around a bit. This is done in the next step.

(6) Execute by hand:

[RCL 0], [STO 1], [RCL 3], [STO 2], [STO + 0],

[RCL 7], [STO 4], [RCL 6], [STO 5], [STO + 7].

(7) Assume you have a program stored, commencing in program register 19, that will calculate $-g(x, y)$, $-h(x, y)$, and $E(x, y)$, putting them respectively in register three, register six, and the display; the calculation is to be done taking x and y from registers zero and seven respectively.

(8) Press GTO. 19 and R/S. This will execute the program referred to in (7).

(9) Now branch, depending on the value of $E(x, y)$ in the display.

(a) If it seems small enough, stop, and read the values of x and y from registers zero and seven respectively.

(b) If it is not small enough, but is smaller than $E(x_r, y_r)$, return to step (1).

(c) If it is greater than or equal to $E(x_r, y_r)$, execute the following by hand:

[2], [STO ÷ 2], [STO ÷ 5],

[RCL 1], [RCL 2], [+], [STO 0],

[RCL 4], [RCL 5], [+], [STO 7].

Then return to step (8).

PROGRAM APPENDIX FOR THE TI-57

Unless specifically stated otherwise, the programs are to be stored as follows. Store the first instruction in program register 00, the next in program register 01, and so on consecutively. Then, to run the program, simply push the R/S key, provided program control is at the beginning of program memory. If it is not, or if you are not sure, press RST first.

Program II. 1. Evaluation of a polynomial by Horner's method.

This program is first referred to in Sect. 1 of Chap. II. It uses Horner's method to evaluate the polynomial, p, defined by

(II. 1. 1) $$p(x) = a_0x^n + a_1x^{n-1} + \ldots + a_{n-1}x + a_n$$

at c. The coefficients $a_0, a_1, \ldots, a_n$ are to be input in the course of storing the program. Start with c in the display.

II. 1 [STO 7],

a_0, [SBR 1],

a_1, [SBR 1],

a_2, [SBR 1],

. . . . . .

a_{n-1}, [SBR 1],

a_n, [=], [R/S], [RST],

[Lbl 1], [=], [×], [RCL 7], [+], [INV SBR].

Program II. 2. Polynomial evaluation

At the end of the program, $p(c)$ will be in the display. Note that the last instruction of the program returns control to the beginning, so that if then you wish to calculate $p(c^*)$ for some c^* other than c, just input c^* and press R/S.

Unless $n = 1$, Program (1. 6) in Chap. II uses as many steps as, or more steps than, Program II. 1, and the latter might better be used.

Program II. 2. Evaluation of a sixth degree polynomial by Horner's method (using stored coefficients).

The coefficients $a_0, a_1, \ldots, a_6$ are to be stored respectively in registers zero, one, ..., six. Start with c in the display

II. 2 [STO 7],

[RCL 0], [SBR 1],

[RCL 1], [SBR 1],

[RCL 2], [SBR 1],

[RCL 3], [SBR 1],

[RCL 4], [SBR 1],

[RCL 5], [SBR 1],

[RCL 6], [=], [R/S], [RST],

[Lbl 1], [=], [×], [RCL 7], [+], [INV SBR].

At the end of the program, $p(c)$ will be in the display.

Suppose you wish to evaluate a fourth degree polynomial. You can artificially make a sixth degree polynomial out of it by adding the terms $0x^6 + 0x^5$. Then Program II. 2 will evaluate it. Alternatively, you can store $a_0, a_1, \ldots, a_4$ respectively in registers zero, one, ..., four. Then replace the program line

[RCL 4], [SBR 1],

by

[RCL 4], [=], [R/S], [RST],

and omit the next two lines, but not the final line.

If you have a polynomial of degree higher than six, you can work in a few lines of Program II. 1 toward the end of Program II. 2.

There are occasions when you might wish to see the successive b_r's that are calculated by Horner's method. See Sect. 2 of Chap. II for a discussion of this. To accomplish this, insert [R/S] between [=] and [×] in the last line of either Program II. 1 or Program II. 2. Now the program will stop every time a b_r appears in the display. After you have done whatever you wish with that b_r, you press R/S, and the program will go on to the next b_r.

Program II. 3. Deflation of a sixth degree polynomial.

Let the coefficients of the polynomial be stored as for Program II. 2. Start with c in the display.

II. 3 [STO 7], [RCL 0],

[SBR 1], [RCL 1], [=], [STO 1],

[SBR 1], [RCL 2], [=], [STO 2],

[SBR 1], [RCL 3], [=], [STO 3],

[SBR 1], [RCL 4], [=], [STO 4],

[SBR 1], [RCL 5], [=], [STO 5],

[SBR 1], [RCL 6], [=], [R/S], [RST],

[Lbl 1], [×], [RCL 7], [+], [INV SBR].

At the end, b_r has replaced a_r for $r = 0, 1, \ldots, 5$, and $b_6 = p(c)$ is in the display. This program is used when one wishes to deflate a polynomial. See Sect. 2 of Chap. II.

Program II. 4. Evaluation of a fifth degree polynomial and its derivative by Horner's method.

The coefficients $a_0, a_1, \ldots, a_5$ are to be stored respectively in registers zero, one,... ; five. Start with c in the display.

II. 4 [STO 6], [×], [RCL 0], [STO 7], [+],

[RCL 1], [SBR 1],

[RCL 2], [SBR 1],

[RCL 3] , [SBR 1] ,

[RCL 4] , [SBR 1] ,

[RCL 5] , [=] , [R/S] , [RST] ,

[Lbl 1] , [=] , [+] , [x ⇄ t] , [×] , [RCL 6] , [=] , [x ⇄ t] , [×] , [RCL 6] ,

[+] , [INV SBR] .

At the end, p(c) is in the display and p'(c) is in register seven. If you press x ⇄ t, then p(c) and p'(c) will change places, bringing p'(c) into the display. Pressing x ⇄ t again will put them back as they were.

If you have a polynomial of degree less than five, proceed as explained after Program II. 2.

Program VII. 1. Bisection.

This is a program to carry out the search for a root by the bisection method. See Sect. 3 of Chap. VII. Start with m_r and h_r respectively in registers zero and one, and with 0 in register seven. If $f(m_r) = 0$, the program stops with zero in the display. Otherwise, it puts m_{r+1} and h_{r+1} respectively in registers zero and one, and stops with h_{r+1} in the display. As h_{r+1} is never zero, you can tell if you have found a zero, or if you must run the program again.

VII. 1 2, [INV Prd 1] , [SBR 1] ,

[x = t] , [R/S] ,

[INV x ≥ t] , [GTO 2] ,

[RCL 1] , [SUM 0] , [R/S] , [RST] ,

[Lbl 2] , [RCL 1] , [INV SUM 0] , [R/S] , [RST] ,

[Lbl 1] , subroutine .

The subroutine depends on what function f you are working with. It should be written to calculate $f(m_r)$, taking m_r from register zero, and to put $f(m_r)$ in the display.

Program VII. 2. Newton's method for polynomial root finding.

This is a program for approximating a zero of a polynomial of degree six. In the first place, we must have $a_0 \neq 0$, since otherwise the polynomial is of degree five or less, and we proceed as in Sect. 5 of Chap. VII. So we divide the polynomial by a_0. This does not change the location of any zeros of the polynomial, but it makes the leading coefficient equal to unity. We will take advantage of this.

We first store the coefficients $a_1, a_2, \ldots, a_6$ respectively in registers zero, one, ..., five. We put a_0 in the display. Then we execute the steps

[INV Prd 0], [INV Prd 1], [INV Prd 2],

[INV Prd 3], [INV Prd 4], [INV Prd 5].

Now, except for the leading coefficient (which is unity), we have stored the coefficients of the polynomial we got by dividing the original polynomial by a_0.

Now we seek a zero by Newton's method (see Sect. 5 of Chap. VII). Let x_r be an approximation to a zero. Put x_r into the display. Next, for the polynomial p that we are now dealing with, we evaluate p and p' both at x_r and stop to see if $p(x_r)$ seems near enough to 0 to satisfy us. If not, we calculate x_{r+1} by (5. 3) of Chap. VII, namely

$$x_{r+1} = x_r - \frac{p(x_r)}{p'(x_r)},$$

with x_{r+1} finally in the display, and stop for inspection. But now x_{r+1} is in the display, and we can just repeat the process. After each stop, we simply press R/S to get going again since the program takes care of getting us to the beginning.

VII. 2 [STO 6], [STO 7], [1], [x⇌t], [+],

[RCL 0], [SBR 1],

[RCL 1], [SBR 1],

[RCL 2], [SBR 1],

[RCL 3], [SBR 1],

[RCL 4], [SBR 1],

[RCL 5], [=], [R/S],

[÷], [x⇌t], [=], [INV SUM 6], [RCL 6], [R/S], [RST],

[Lbl 1], [=], [+], [x⇌t], [×], [RCL 6], [=], [x⇌t], [×],

[RCL 6], [+], [INV SBR].

Program XIII. 1. Damped Newton's method for finding simultaneous solutions of two equations.

The theory for this is given in Sect. 5 of Chap. XIII. We seek points (x, y) such that

$$g(x, y) = 0, \qquad h(x, y) = 0$$

are both true. We have a point (x_r, y_r) which we hope is "close" to a simultaneous solution, and wish to find (x_{r+1}, y_{r+1}) which is "closer." To monitor the closeness, we define

$$E(x, y) = (g(x, y))^2 + (h(x, y))^2 .$$

So we seek (x_{r+1}, y_{r+1}) for which

$$E(x_{r+1}, y_{r+1}) < E(x_r, y_r) .$$

Parts of the program have to be done by pressing keys by hand. Also, we have to keep a written record of $E(x_r, y_r)$, $E(x_{r+1}, y_{r+1}), \ldots$.

(1) We start with x_r, y_r, $-g(x_r, y_r)$, $-h(x_r, y_r)$ respectively stored in registers zero, seven, three, and six and with $E(x_r, y_r)$ in the display.

(2) Write down $E(x_r, y_r)$.

(3) By hand, calculate $g_x(x_r, y_r)$, $g_y(x_r, y_r)$, $h_x(x_r, y_r)$, $h_y(x_r, y_r)$ and store them respectively in registers one, two, four, and five.

(4) We assume that you have stored Program (3. 7), AE, of Chap. XIII, with [R/S] added to the end, in program registers 00 through 20, but with the following modifications: replace [STO ÷ 2], [STO ÷ 3], [STO − 5], [STO − 6], [STO ÷ 6], [STO − 3] respectively by [INV Prd 2], [INV Prd 3], [INV SUM 5], [INV SUM 6], [INV Prd 6], [INV SUM 3].

(5) Press RST and R/S. This runs modified Program (3. 7), AE, of Chap. XIII. This solves the equations (4. 4) of Chap. XIII, and there are now numbers δx and δy respectively in registers three and six. These are proposed corrections for x_r and y_r, so that we should consider taking

$$x_{r+1} = x_r + \delta x, \qquad y_{r+1} = y_r + \delta y .$$

But first we have to check if this will decrease E. This requires shuffling $x_r, y_r, \delta x$, and δy around a bit. This is done in the next step.

(6) Execute by hand:

[RCL 0], [STO 1], [RCL 3], [STO 2], [SUM 0],

[RCL 7], [STO 4], [RCL 6], [STO 5], [SUM 7].

(7) Assume you have a program stored, commencing with [Lbl 1] and beginning in any program register after 20, that will calculate $-g(x, y)$, $-h(x, y)$, and $E(x, y)$, putting them respectively in register three, register six, and the display; the calculation is to be done taking x and y from registers zero and seven respectively.

(8) Press GTO 1 and R/S. This will execute the program referred to in (7).

(9) Now branch, depending on the value of $E(x, y)$ in the display.

(a) If it seems small enough, stop, and read the values of x and y from registers zero and seven respectively.

(b) If it is not small enough, but is smaller than $E(x_r, y_r)$, return to step (1).

(c) If it is greater than or equal to $E(x_r, y_r)$, execute the following by hand:

[2], [INV Prd 2], [INV Prd 5],

[RCL 1], [+], [RCL 2], [=], [STO 0],

[RCL 4], [+], [RCL 5], [=], [STO 7].

Then return to step (8).

ANSWERS FOR SELECTED PROBLEMS

CHAPTER I

2\. 4, pp. 13-14.

(a) RPN: x, [↑], [x^2], [sin], [x⇄y], [cos], [x^2], 1, [+], [÷].

(a) AE: x, [STO], [x^2], [sin], [÷], [(], [RCL], [cos], [x^2], [+], 1, [=].

(b) RPN: x, [↑], 1, [−], [↑], [↑], 2, [+], [÷], $\sqrt{x}$.

(b) AE: x, [−], 1, [=], [STO], [÷], [(], [RCL], [+], 2, [=], $\sqrt{x}$.

(c) Note that $f(x) = 1 - \sqrt{(x-1)^2}$. Take it from there.

4\. 1, p. 16. 1.940075 by interpolation as against 1.939635 given by the e^x key.

CHAPTER II

1\. 1, p. 22.

c	p(c)	c	p(c)
-3	2179 16644	1	100
-2	1 16281	2	10 46529
-1	0	3	8716 66576
0	1		

1\. 3, p. 23.

c	p(c)	c	p(c)
-2	65	3	-20
-1	12	4	17
0	1	5	156
1	-4	6	481
2	-15		

1\. 4, p. 23. p(c) = 0 for c = 0.3 and c = 3.7, approximately. So approximately, p(x) is negative for $0.3 < x < 3.7$, and positive otherwise.

1\. 5, p. 24.

c	p(c)	c	p(c)
-2	181	1	-5
-1	1	2	1
0	-1	3	221

2\. 1, p. 27-28. (a), -2; (b), 7; (c), -1/3; (d), 3; (e), 1.

2. 2, p. 28. Limit = $n/2$.
2. 3, p. 28. $p(x) = (x-4)(x^3+2x+4) + 17$.
2. 4, p. 28. $p(x) = (x+1)(8x^3 - 22x^2 + 13x - 2) + 1$.
$p(x) = (x-3)(8x^3 + 10x^2 + 21x + 74) + 221$.
2. 9, p. 30. $q(x) = 8x^2 - 6x + 1$.
3. 3, p. 38.

c	p(c)	p'(c)
0	1	- 4
1	- 4	- 8
2	-15	-12
3	-20	8
4	17	76

CHAPTER III

1. 1, p. 40. 1. 3, -1/3; 2. -1/3, 3; 3. -1, 1; 4. 0, ∞; 5. 0, ∞; 6. ∞, 0; 7. 2, -1/2; 8. 2/3, -3/2.
1. 2, p. 40. 1. Rectangle; 2. Rectangle; 3. No; 4. No.
1. 3, pp. 40-41. 1. Yes; 2. No; 3. No.
1. 4, p. 41. 1. $2y = 3x$; 2. $y = 1$; 3. $x = 1$; 4. $3x + 4y + 2 = 0$, 5. $x = -2$; 6. $x + y = 4$; 7. $y = 0$; 8. $x = 0$; 9. $x + y = 1$.
1. 5, p. 41. (a), $x - 2y + 6 = 0$; (b), 2/5, 16/5; (c), $6/\sqrt{5} \cong 2.6832\ 81574$.
1. 6, p. 42.

Δx	$(\sqrt{4+\Delta x}-\sqrt{4})/\Delta x$	Δx	$(\sqrt{4+\Delta x}-\sqrt{4})/\Delta x$
1	0. 23606 79770	-1	0. 26794 91920
0. 1	0. 24845 67300	-0. 1	0. 25158 23400
0. 01	0. 24984 39000	-0. 01	0. 25015 64000
0. 001	0. 24998 40000	-0. 001	0. 25001 60000
0. 0001	0. 25000 00000	-0. 0001	0. 25000 00000
0. 00001	0. 25000 00000	-0. 00001	0. 25000 00000

1. 7, p. 43.

Δx	$(\sin(\frac{\pi}{3}+\Delta x)-\sin(\frac{\pi}{3}))/\Delta x$	Δx	$(\sin(\frac{\pi}{3}+\Delta x)-\sin(\frac{\pi}{3}))/\Delta x$
1	0. 02262 56114	-1	0. 81884 53737
0. 1	0. 45590 18850	-0. 1	0. 54243 22810
0. 01	0. 49566 15700	-0. 01	0. 50432 17600
0. 001	0. 49956 69000	-0. 001	0. 50043 29000
0. 0001	0. 49995 70000	-0. 0001	0. 50004 30000
0. 00001	0. 49999 00000	-0. 00001	0. 50001 00000

1. 8, p. 43.

Δx	$(f(1+\Delta x)-f(1))/\Delta x$	Δx	$(f(1+\Delta x)-f(1))/\Delta x$
0. 1	0. 23809 52380	-0. 1	0. 26315 78950
0. 01	0. 24875 62200	-0. 01	0. 25125 62800
0. 001	0. 24987 51000	-0. 001	0. 25012 51000
0. 0001	0. 24998 80000	-0. 0001	0. 25001 30000
0. 00001	0. 25000 00000	-0. 00001	0. 25000 00000

1. 9, p. 43. A = 0. 05, B = 1/3. 35 ≅ 0. 29850 74627.
2. 1, p. 47.

(a) $$\frac{(x+\Delta x)^2 - x^2}{\Delta x} = 2x + \Delta x\ .$$

(b) $$\frac{(x+\Delta x)^4 - x^4}{\Delta x} = 4x^3 + 6x^2\Delta x + 4x(\Delta x)^2 + (\Delta x)^3\ .$$

(c 1) $$\frac{g(x+\Delta x)h(x+\Delta x)-g(x)h(x)}{\Delta x} = g(x+\Delta x)\frac{h(x+\Delta x)-h(x)}{\Delta x}+h(x)\frac{g(x+\Delta x)-g(x)}{\Delta x}.$$

(c 2) $$\frac{g(h(x+\Delta x))-g(h(x))}{\Delta x} = \frac{g(h(x+\Delta x)-g(h(x))}{h(x+\Delta x)-h(x)} \times \frac{h(x+\Delta x)-h(x)}{\Delta x}.$$

CHAPTER IV

1. 1, pp. 49-50. $2/3 \cong 0.66666\ 66667$; $\ln 2 \cong 0.69314\ 71806$; $\pi \cong 3.1415\ 92654$; $\sqrt{2} \cong 1.4142\ 13562$; $\sin 1 \cong 0.84147\ 09848$; $2^{49} \cong 5.6294\ 99534 \times 10^{14}$; $\tan(1.5612\ 91773) \cong 105.20\ 95563$.

1. 2, p. 50. $B = 0.234951$ will do.

1. 6, p. 58. Its derivative is $n(f(x))^{n-1}f'(x)$, which is liable to be large if n is large.

1. 7, p. 58. We get

$$\begin{array}{r} \pi/2 \cong 1.5707\ \ 96326\ \ 79490 \\ 1.5612\ \ 91773 \\ \hline d \cong 0.0095\ \ 04553\ \ 79490 \end{array}$$

By calculator

$$\frac{1}{\tan(0.00\ \ 95045\ \ 53795} \cong 105.20\ \ 95563$$

which agrees to 10 digits with the correct answer.

1. 8, p. 58. We get

$$\begin{array}{r} \pi/2 \cong 1.5707\ \ 96326\ \ 79490 \\ 1.104\sqrt{2} \cong 1.5612\ \ 91772\ \ 85989 \\ \hline D \cong 0.0095\ \ 04553\ \ 93501 \end{array}$$

By calculator

$$\frac{1}{\tan(0.00\ \ 95045\ \ 53935} \cong 105.20\ \ 95548,$$

which is a 10-digit rounded approximation to the right side of (1. 8) of Chap. IV.

2. 2, p. 60,
2. 3, p. 60,
2. 5, p. 60,
2. 6, p. 60-61.

} Are to be compared respectively with accurate answers that you can get by your answers to 3. 1, p. 64, 3. 2, p. 64, 3. 4, p. 65, and 3. 8, p. 66.

2. 4, p. 60. The calculator gives

$$\ln(3.1462) + 2 \cong 3.1461\ 95375.$$

If this is subtracted from $x = 3.1462$, the first six digits are cancelled out. The answer cannot be accurate to more than 4 digits on a 10 digit calculator.

2. 7, p. 62. On the HP-33E, we get

$$p_{50}(-10) = 4.5255\ \ 00000 \times 10^{-5}$$

However, rounded to six significant digits, we have $e^{-10} \cong 4.5399\ \ 9 \times 10^{-5}$. So we got only the first two digits correct.

2. 8, p. 62. Because you did not use exceptionally accurate values for $\pi/2$ and $1.104\sqrt{2}$, the cancellation will cause you to get an inaccurate value for D, and everything is inaccurate from there on.

3. 1, p. 64. $$\frac{\sqrt{x+\Delta x}-\sqrt{x}}{\Delta x} \times \frac{\sqrt{x+\Delta x}+\sqrt{x}}{\sqrt{x+\Delta x}+\sqrt{x}} = \frac{1}{\sqrt{x+\Delta x}+\sqrt{x}}.$$

3. 2, p. 64. By the hint, we get

$$\frac{\sqrt[3]{x+\Delta x}-\sqrt[3]{x}}{\Delta x}=\frac{1}{(x+\Delta x)^{\frac{2}{3}}+(x+\Delta x)^{\frac{1}{3}}(\Delta x)^{\frac{1}{3}}+(\Delta x)^{\frac{2}{3}}}.$$

3. 3, p. 64. 683. 00 14641 and 1. 4641 25705 $\times 10^{-3}$.

3. 4, p. 65. $\frac{\sin(x+\Delta x)-\sin(x-\Delta x)}{2\Delta x}=\frac{(\cos x)(\sin \Delta x)}{\Delta x}$.

3. 5, p. 65. $\frac{\cos(\alpha+\Delta x)-\cos(\alpha-\Delta x)}{2\Delta x}=\frac{(\sin \alpha)(\sin \Delta x)}{\Delta x}$.

3. 6, p. 65. $1-\cos x=(\sin^2 x)/(1+\cos x)$.

3. 7, p. 65. We get on the HP-33E, $p_{43}(10)\cong 22026.\ 465\ 82$. Taking the reciprocal gives $e^{-10}\cong 4.\ 5399\ \ 92971\times 10^{-5}$. True value, rounded to 10 digits, is $4.\ 5399\ \ 92976\times 10^{-5}$.

3. 8, p. 66. For small h, (3. 10) of Chap IV gives $\sinh h\cong h\cosh h$ as a good approximation.

3. 9, p. 70. We get

$$x^4-8x^3+12x^2+16x+4$$
$$=(x-4.45)(x^3-3.55x^2-3.7975x-0.898875)+6.25\times 10^{-6}.$$

Putting $x=40/9$ in this gives 0. 000 60966 32588.

3. 10, p. 70. Instead of (3. 19) and (3. 20), you get

$$h^2+k^2+1.4142h+1.4142k-1.918\times 10^{-5}$$
$$2hk+1.4142h+1.4142k-1.918\times 10^{-5}.$$

CHAPTER V

2. 3, p. 79. See figure below.

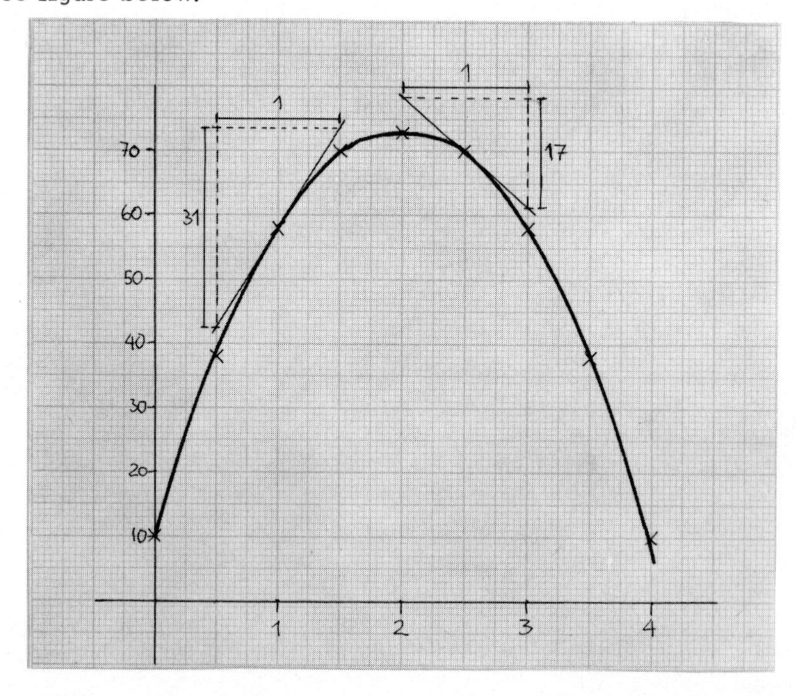

ANSWERS

2.4, p. 79. Using (2.3), 24, -24, and -8. Using (2.4), 32, -16, and 0.

3.1, p. 80. $p'(2) \cong -103.28\ 68175$. With $h = 10^{-1}$, (2.2) gives $-120.61\ 11480$; error $\cong 17.32$. With $h = 10^{-2}$, (2.2) gives $-103.43\ 61210$; error $\cong 0.1493$. With $h = 10^{-3}$, (2.2) gives $-103.28\ 83150$; error $\cong 0.0014975$.

CHAPTER VI

3.1, p. 92. Let $f(x) = \ln x$. Then $f''(x) = -1/x^2$. For $2 \leq x \leq 3$, the largest this can be in absolute value is $1/4$. With arguments 0.01 apart, $(x_1-x_2)^2 = 10^{-4}$. So the maximum error would be $(1/4)(1/8)10^{-4} = 0.00000\ 3125$. So linear interpolation would give about 5-digit accuracy.

CHAPTER VII

2.1, p. 102. $3.4 \leq x \leq 3.5$.

2.3, p. 103. Two zeros.

3.1, p. 108. $c \cong 0.15859$.

3.2, p. 108-109. $c \cong 0.26795$.

5.2, p. 118. $c \cong 4.5052\ 41496$.

5.6, p. 119. $c \cong -0.25992\ 10499$.

5.7, pp. 119-220. $8x^4 - 14x^3 - 9x^2 + 11x - 1 = 8(x-c_1)(x-c_2)(x-c_3)(x-c_4)$, where $c_1 \cong -0.97682\ 35894$, $c_2 \cong 0.10036\ 33318$, $c_3 \cong 0.64274\ 66710$, $c_4 \cong 1.9837\ 13586$.

5.8, p. 120. $x^4 - 3x^3 - 4x - 1 = (x-c_1)(x-c_2)(x^2 + ax + b)$, where $c_1 \cong -0.23895\ 21590$, $c_2 \cong 3.3767\ 69470$, $a \cong 0.13781\ 73110$, $b \cong 1.2393\ 31900$.

5.9, p. 120. $x^3 - 3x^2 + x - 1 = (x-c)(x^2+ax+b)$, where $c \cong 2.7692\ 92354$, $a \cong -0.32070\ 76460$, $b \cong 3.6110\ 30799$.

5.10, p. 120. $x^3 + 12x - 4 = (x-c)(x^2 + ax + b)$ where $c \cong 0.33032\ 96009$, $a \cong 0.33032\ 96009$, $b = 12.109\ 11765$.

6.3, p. 122. $x_{r+1} = \{2x_r + A/x_r^2\}/3$.

7.1, p. 125. According to the theory in Sect. 7-6 of T-F, if p is a polynomial of degree less than 4, then

$$\int \frac{p(x)dx}{(x-c_1)(x-c_2)(x-c_3)(x-c_4)}$$

$$A_1 \ln(x-c_1) + A_2 \ln(x-c_2) + A_3 \ln(x-c_3) + A_4 \ln(x-c_4) + C,$$

if the c_i are all different. Differentiate the above, and multiply both sides of the equation by

$$(x-c_1)(x-c_2)(x-c_3)(x-c_4).$$

Set $x = c_1$. This will give a formula for A_1 in terms of $p(c_1)$ and the c_i. Write a program to evaluate this formula. By rearranging the c_i's, the same program will give the values of A_2, A_3, and A_4. Indeed, you might as well write the program to rearrange the c_i's, so that if you run the program four times, you will get A_1, A_2, A_3, A_4 in order. Check the program by taking c_1, c_2, c_3, c_4 respectively to be 1, 2, 3, 4 for which the values of A_1, A_2, A_3, A_4 can easily be got. Now

use the values for the c_i's from Prob. 5. 7, pp. 119-120, and get
$A_1 \cong -0.18025\ 69377$, $A_2 \cong -0.99086\ 32788$, $A_3 \cong 1.0438\ 73611$,
$A_4 \cong 0.12724\ 66072$.

7. 2, p. 125. According to the theory in Sect. 7-6 of T-F, if p is a polynomial of degree less than 4, then

$$\int \frac{p(x)\,dx}{(x-c_1)(x-c_2)(x^2+ax+b)}$$
$$= A_1 \ln(x-c_1) + A_2 \ln(x-c_1) + A_3 \ln(x^2+ax+b) + A_4 \tan^{-1} d(x+\tfrac{a}{2}) + C\,,$$

where

$$d = \frac{2}{\sqrt{4b-a^2}}\,,$$

if $c_1 \neq c_2$ and $4b - a^2 > 0$. Differentiate the above, and multiply both sides of the equation by

$$(x-c_1)(x-c_2)(x^2+ax+b)\,.$$

Call this the auxiliary equation. In it, put $x = c_1$. This will give a formula for A_1. Write a program to evaluate this formula. Interchange c_1 and c_2, and the same program will give A_2. Now that you know A_1 and A_2, set $x = -a/2$ in the auxiliary equation, and get a formula for A_4. Finally take x to be any fourth number in the auxiliary equation, and get a formula for A_3 (involving A_1, A_2 and A_4, but you know them). Check the program by using some "easy" values for c_1, c_2, a, and b. Now use the values of c_1, c_2, a, and b from Prob. 5. 8, p. 120, and get

$$A_1 = -0.21889\ 22344,\quad A_2 = 0.0\ 21100\ 48291$$
$$A_3 = 0.0\ 98895\ 87493,\quad A_4 = -0.0\ 98933\ 38085\,.$$

8. 2, p. 127. $e^{\rho x}$ is a solution for $\rho \cong -0.23895\ 21590$ or $\rho \cong 3.3767\ 69470$. Also for any values of A and B, $e^{(-0.06890\ 86555x)}\{A\cos(1.1111\ 18130)x + B\sin(1.1111\ 18130)x\}$ is a solution.

CHAPTER VIII

1. 4, p. 136. $2\pi \cong 6.2831\ 85308$.
2. 5, p. 140. }
2. 6, p. 140. } See answer for 3. 2, p. 145.
3. 2, p. 145. For $x = \ln x + 2$, we get

r	x_r	x^*	r	x_r	x^*
1	1. 0000 00000		7	3. 1410 17985	3. 1462 55356
2	2. 0000 00000		8	3. 1445 46946	3. 1461 99482
3	2. 6931 47181	4. 2588 91358	9	3. 1456 69824	3. 1461 93852
4	2. 9907 10466	3. 2145 41389	10	3. 1460 26848	3. 1461 93285
5	3. 0955 10973	3. 1524 88499	11	3. 1461 40339	3. 1461 93228
6	3. 1299 52989	3. 1468 13107	12	3. 1461 76412	3. 1461 93220

3. 2, p. 145. For $x = e^{x-2}$, we get

r	x_r	x^*	r	x_r	x^*
1	1. 00000 00000		6	0. 15874 48861	0. 15859 38047
2	0. 36787 94412		7	0. 15861 82172	0. 15859 43261
3	0. 19551 45341	0. 13089 39882	8	0. 15859 81264	0. 15859 43391
4	0. 16455 91061	0. 15778 27675	9	0. 15859 49401	0. 15859 43395
5	0. 15954 31446	0. 15857 31991	10	0. 15859 44348	0. 15859 43396

3. 3, p. 148. Taking four triples in succession, we get: 15. 30, 15. 30, 15. 30, 15. 30.

CHAPTER IX

1. 2, p. 151. Local minimum at $x = 1$, function value = 0; local minimum at $x = 10$, function value = $(\ln 10)/10 \cong 0.23025\ 85093$; absolute maximum at $x = e$, function value = $1/e \cong 0.36787\ 94412$.

1. 3, p. 152. Maximum area for $\theta = \cos^{-1}\{(\sqrt{33} - 1)/8\} \cong 0.93592\ 94556^r$
$\cong 53.624\ 80733^o$.

Maximum area $\cong 3.5203\ 45186$.

CHAPTER X

2. 2, p. 167. 1819. 1 459 by the antiderivative. 1819. 1 161 by Simpson's rule.

3. 1, p. 170.

N	M(h)	error $(h)/h^2$	M(h)+ correction	error $(h)/h^4$
1	0. 66666 66667	0. 026480	0. 69791 66667	4.7695×10^{-3}
2	0. 68571 42855	0. 029731	0. 69352 67855	6.0737×10^{-3}
4	0. 69121 98913	0. 030836	0. 69317 30163	6.6139×10^{-3}
8	0. 69266 05541	0. 031144	0. 69314 88354	6.7781×10^{-3}
16	0. 69314 71806	0. 031223	0. 69314 72847	6.8223×10^{-3}

(i) See table; (i) Smaller, since $f''(\xi) > 0$ for $f(x) = 1/x$ and $\xi > 0$; (iii) From table, $K \cong 0.0313$; (iv) $N = 80$; (v) $N = 25020$.

3. 2, p. 170. Max $|f''(\xi)| = 2$, $N = 130\,(40825)$ for 5 (10) place accuracy.

3. 3, p. 171. See table in answer for problem 3. 1, p. 170.

3. 4, p. 173.

N	T(h)	error $(h)/h^2$	T(h)+ correction	error $(h)/h^4$
1	0. 75	-5.6852×10^{-2}	0. 6875	5.6472×10^{-3}
2	0. 70833 33335	-6.0745×10^{-2}	0. 69210 83333	7.0216×10^{-3}
4	0. 69702 38095	-6.2026×10^{-2}	0. 69311 75595	7.5830×10^{-3}
8	0. 69412 18503	-6.2379×10^{-2}	0. 69314 52878	7.7529×10^{-3}
16	0. 69339 12019	-6.2469×10^{-2}	0. 69314 70613	7.8184×10^{-3}

(i) See table; (ii) larger, since $f''(\xi) > 0$ for $f(x) = 1/x$ and any $\xi > 0$; (iii) See table, $K \cong -0.0625$; (iv) $N = 112$; (v) $N = 35356$.

3. 5, p. 174. See table for answer to problem 3. 4, p. 173.

4. 1, p. 176-177.

N	S(h)	error(h)/h^4
1	0. 69444 44443	-1.2972×10^{-3}
2	0. 69325 39683	-1.7086×10^{-3}
4	0. 69315 45310	-1.8817×10^{-3}
8	0. 69314 76527	-1.9337×10^{-3}
16	0. 69314 72103	-1.9464×10^{-3}

(i) Calculated from answers to 3. 1, p. 170, and 3. 4, p. 173, see table; (ii) larger, since $f^{iv}(\xi) > 0$ for $f(x) = 1/x$ and any $\xi > 0$; (iii) see table, $K \cong -0.00195$; (iv) $M = 10$; (v) $M = 160$.

4. 2, p. 177. $\max |f^{iv}(\xi)| = 24$; $M = 14(228)$ for 5(10) place accuracy.

4. 3, p. 177. Since $f^{iv} \equiv 0$ in case f is a polynomial of degree ≤ 3, hence (4. 3) then vanishes.

4. 4, p. 177. $M(5) = 21.875$, $T(5) = -87.5$, so $S(2.5) = -14\frac{7}{12}$.

CHAPTER XI

1. 5, p. 182. Ground speed 345. 01 28324 knots 172. 87 908118° West from North.

1. 6, p. 183.
 (a) (0. 70710 67812, -0. 70710 67812).
 (b) (-0. 70710 67812, -3. 5355 33905).
 (c) (-6. 3639 61035, 7. 7781 74588).
 (d) (-2. 8284 27124, -1. 4142 13563).

3. 5, p. 189.
$\tan^{-1}(5/3) \cong 59.036\ 24347^\circ$
$\tan^{-1} 5 \cong 78.690\ 06753^\circ$
$\tan^{-1}(10/11) \cong 42.273\ 68901^\circ$
Sum $\cong 180.00\ 00000^\circ$.

3. 8, p. 190. 5.

4. 4, p. 194. $\vec{F} = \frac{106}{50}\vec{i} - \frac{5}{50}\vec{j} - \frac{133}{50}\vec{k}$.

4. 7, p. 196.
$A = \cos^{-1}(\sqrt{2}/\sqrt{19}) \cong 71.068\ 17682$
$B = \cos^{-1}(15/19) \cong 37.863\ 64636$
$C = \cos^{-1}(\sqrt{2}/\sqrt{19}) \cong 71.068\ 17682$
Sum $\cong 180.00\ 00000$.

4. 8, p. 196. $\cos^{-1}(1/\sqrt{3}) \cong 54.735\ 61033$

4. 9, p. 196. $\cos^{-1}(\sqrt{6}/3) \cong 35.264\ 38968$

5. 1, p. 198. Since $\vec{F}$ is in the plane of $\vec{A}$ and $\vec{B}$, it must be perpendicular to $\vec{A} \times \vec{B}$. But it was also required to be perpendicular to $\vec{B}$. However, the vector $(\vec{A} \times \vec{B}) \times \vec{B}$ is perpendicular to both $\vec{A} \times \vec{B}$ and $\vec{B}$.

5. 2, p. 200. $-\vec{i} - 3\vec{j} + 4\vec{k}$.

5. 3, p. 200. $c(2\vec{i} + \vec{j} + \vec{k})$, with c a scalar.

5. 4, p. 200. $2\sqrt{6} \cong 4.8989\ 79486$.

5. 5, p. 200. $\sqrt{6}/2 \cong 1.2247\ 44872$.

5. 6, p. 200. $11/\sqrt{107} \cong 1.0634\ 10138$.

5. 7, p. 200. $7x - 5y - 4z = 6$.

5. 8, pp. 200-201. $17x - 26y + 11z = -40$.

6. 1, p. 201. 2/3.

CHAPTER XII

2. 3, p. 211. $\frac{47}{24}$ cubic feet. Approximately 1. 9583 33333 .

2. 4, p. 211. 2.0002 8935 cubic feet.

CHAPTER XIII

1. 1, p. 215. High: (2/3, 4/3, 0).

1. 2, p. 215. Low: (0, 0, 0); high: $(\pm 1, \pm 1, \sqrt{2})$. The partial derivatives do not exist at (0, 0). They exist but are not zero at $(\pm 1, \pm 1)$.

2. 2, p. 219. 0. 2631 06796.

2. 3, p. 219. The average of the y_n's.

2. 4, p. 219. y = 0. 123x + 3. 580.

3. 1, p. 221. (a) x = 1. 2, y = 0. 9.

(b) x = 0. 4, y = 1. 3. There is an error message because $a_{11} = 0$, and so division by a_{11} is not permitted. Interchange the two equations, and all goes well.

(c) There is no solution. There is an error message because $a'_{22} = 0$. There is no way to get out of this, because it is not possible to get a solution of the pair of equations. If you multiply the second equation by 3, you would be trying to get 30x + 60y equal to both 78 and 90 .

4. 1, p. 224. (a) Critical point is (3. 6). The early entries in the tables for (b) and (c) have been rounded, to save writing

(b)

r	x_r	y_r	$g(x_r, y_r)$	$h(x_r, y_r)$
0	0. 5	0	-0. 5	-1
1	0. 66667	-0. 25	-0. 064815	0. 52083
2	0. 65186	-0. 17811	-0. 018930	0. 024017
3	0. 65434 88	-0. 17372 74	-2.85×10^{-5}	-2.95×10^{-5}
4	0. 65435 52613 7	-0. 17373 07240 6	2×10^{-10}	2×10^{-10}
5	0. 65435 52613 3	-0. 17373 07240 4	1×10^{-10}	0

(c)

r	x_r	y_r	$g(x_r, y_r)$	$h(x_r, y_r)$
0	7	19	-73	18
1	7. 1362	19. 047	1. 5593	-0. 04421
2	7. 1291 648	19. 028 596	2.9384×10^{-3}	-4.25×10^{-5}
3	7. 1291 40940 5	19. 028 56003 6	-2×10^{-7}	0

5. 1, p. 229. Need two cut downs at first step. After that, convergence in 4 steps to (3. 7568 34008, 2. 7798 57510) with $E = 10^{-18}$.

5. 2, p. 229. Starting, e. g. , with (1, -1), need one cut down at first step. After that, convergence in 4 steps to (1. 3734 78353, -1. 5249 64836), with $E = 10^{-18}$.

CHAPTER XIV

2. 1, p. 232. From (2. 5), $|\text{error}(N)| = 3.1 \times 10^{-6} < 5 \times 10^{-6}$ for N = 3. Evaluating (2. 6) at x = 1 gives 0. 94608 27665.

2.2, p. 232. Substitute y^2 for x in the series
$\sin x = x - x^3/3! + x^5/5! - x^7/7! + \ldots - (-1)^n x^{2n-1}/(2n-1)! + R_n(x)$,
with $|R_n(x)| \leq |x|^{2n+1}/(2n+1)!$. Then integrate.

2.3, p. 232. $\dfrac{|x|^{4N+7}}{(4N+7)((2N+3)!)}$ if $|x| \leq \sqrt{N}$.

2.4, p. 233. $(((\frac{1}{15}\frac{(-x^4)}{7\cdot 6} + \frac{1}{11})\frac{(-x^4)}{5\cdot 4} + \frac{1}{7})\frac{(-x^4)}{2} + 1)\frac{x^3}{3}$.

2.5, p. 233. From 2.3, the expression in 2.4, p. 233, evaluated at $x = 1$, should be accurate enough. Its value is 0.31026 81578, in error by 1.439×10^{-7}.

CHAPTER XV

1.2, p. 240.

h	y_N	error
1	2.5	-8.5
0.5	3.63	-9.63
0.25	6.053	-12.053
0.125	12.744	-18.744

2.4, p. 246.

$\Delta t = 400$		$\Delta t = 200$	
t	H	t	H
0	70	0	70
400	67.43	200	68.67
800	65.16	400	67.44
1200	63.06	600	66.28
1600	61.07	800	65.17
2000	59.13	1000	64.11
		1200	63.08
		1400	62.07
		1600	61.08
		1800	60.10989
		2000	59.14377
Interpolation gives $t \cong 1821$ when $H = 60$		Interpolation gives $t \cong 1811.4$ when $H = 60$	

2.4, p. 246, continued.

$\Delta t = 100$			
t	H	t	H
100	69.32	1100	63.59
200	68.67	1200	63.08
300	68.04	1300	62.58
400	67.44	1400	62.08
500	66.85	1500	61.58
600	66.28	1600	61.09
700	65.72	1700	60.60
800	65.17	1800	60.11359
900	64.64	1900	59.62989
1000	64.11	2000	59.14772
Interpolation gives $t \cong 1823.48$ when $H = 60$			

2.5, p. 247. 1819.1161.

3.1, p. 251.

h	y_N	error
1	26.17	1.83
0.5	27.696	0.304
0.25	27.9678	0.0322
0.125	27.99738	0.00272

4.1, p. 254.

h	y_N	error	y'_N	error
1	-0.36065	-0.30187	0.27870	-0.24690
0.5	-0.60260	-0.05992	0.06646	-0.03465
0.25	-0.64956	-0.01295	0.03737	-0.00556
0.125	-0.65952	-0.00299	0.03278	-0.00098

INDEX

INDEX

INDEX

List of calculator keys, real and symbolic, used in this text.

arithmetic keys

[+] , [−] , [×] , [÷] : add, subtract, multiply, divide

exponentiation key

[y^x]

function keys

replace the number x in the display by the number f(x). The function f is usually written on the key.

[CLX] = [CLR] : f(x) = 0

[CHS] = [+/−] : f(x) = − x

[1/x] , [x^2] = [INV $\sqrt{x}$] , [$\sqrt{x}$] = [INV x^2] , [e^x] = [INV ln x] ,

[ln x] = [INV e^x] , [SIN] , [COS] , [TAN] , [SIN^{-1}] = [INV SIN] ,

[COS^{-1}] = [INV COS] , [TAN^{-1}] = [INV TAN]

[INT] : f(x) = integral part of x

register keys

[STO] c : store the quantity c somewhere

[RCL] c : recall the quantity c from wherever it was stored

[STO n] : store the content of the display in register n

[RCL n] : recall the content of register n into the display

[STO + n] , [STO − n] , [STO × n] , [STO ÷ n] : with x in the display and N in register n, store N+x, N−x, N×x, N/x in register n.

control keys

[R/S] : stop or start a program

specific RPN keys

[↑] : enter into stack (pushing up the stack)